DeWALT®

W9-DGD-697

SPANISH/ENGLISH CONSTRUCTION DICTIONARY
ILLUSTRATED EDITION
PROFESSIONAL REFERENCE

DICCIONARIO DEL CONSTRUCTOR — ESPAÑOL/INGLÉS
EDICIÓN ILUSTRADA
REFERENCIA PROFESIONAL

Paul Rosenberg

Created exclusively
for DeWALT by:

Un producto exclusivo
para DeWALT, hecho por:

PAL
publications®

www.dewalt.com/guides

OTHER TITLES AVAILABLE

Trade Reference Series
- Blueprint Reading
- Construction
- Construction Estimating
- Construction Safety/OSHA
- Datacom
- Electric Motor
- Electrical
- Electrical Estimating
- HVAC/R – Master Edition
- Lighting & Maintenance
- Plumbing
- Residential Remodeling & Repair
- Security, Sound & Video
- Wiring Diagrams

Exam and Certification Series
- Building Contractor's Licensing Exam Guid
- Electrical Licensing Exam Guide
- HVAC Technician Certification Exam Guid
- Plumbing Licensing Exam Guide

For a complete list of The DEWALT Profe
Reference Series visit **www.dewalt.com**

**This book belongs to
Este libro pertenece a:**

Name/Nombre:_____

Company/Compañía: _____

Title/Título:_____

Department/Departamento: _____

Company Address/Dirección de la compañía:

_____

_____

Company Phone/Teléfono de la compañía:

_____

Home Phone/Teléfono particular:

_____

Pal Publications, Inc.
800 Heritage Drive, Suite 810
Pottstown, PA 19464

800-246-2175

ISBN 0-9770003-9-7

2 3 4 5 6 11 10 09 08

Printed in Canada

Pal Publications
800 Heritage Drive, Suite 810
Pottstown, PA 19464
800-246-2175

AVISO DE DERECHOS

AVISO LEGAL

ISBN 0-9770003-9-7

2 3 4 5 6 11 10 09 08

Impreso en Canadá

Preface

It would be ideal for anyone concerned with communicating in both Spanish and English to take the appropriate language courses. In real life, however, that is a very hard thing to do. Most construction professionals have barely enough time to deal with their existing schedules, much less add a course several nights per week. So, while I do recommend that you take the language courses, this book was put together assuming that you have not done so. And I should add that most language courses do not cover construction terms. I strongly recommend that you try to communicate in your opposite language as much as possible. That will make you much better at putting words into sentences, and at getting ideas from your mind into your coworker's mind. Talk to the guy who speaks the other language about lunch, about sports, and so on. (You might even make a friend.)

In addition to its coverage of construction, this book also covers other necessary subjects, such as employment. This is one of the key areas where Spanish- and English speakers first interact, and it is very important to the people actually working (or hoping to work) in the construction business. You will also find material on the construction business, housekeeping, materials, equipment, conversion factors, and measurement. I also added a section on demolition work, which is often the first construction employment for native Spanish-speakers.

I suspect that there are important terms that I have inadvertently left out of this book. After all, construction terminology in either language could easily fill-up several books. I generally defaulted to Mexican terms when there was a difference, although I tried to list alternatives. So, I was forced to make a lot of choices, and I remain quite open to suggestions from readers regarding alternate choices, spellings, and so on. I will update this book on a continual basis and will attempt to include material suggested by readers, and to keep pace with developments in the construction business.

Paul Rosenberg

Prefacio

Sería ideal para cualquier persona interesada en la comunicación tanto en español como en inglés, aprender los dos idiomas. En la vida real, sin embargo, esto es algo difícil de lograr. La mayoría de los profesionales de la construcción disponen escasamente del tiempo necesario para manejar sus horarios actuales y mucho menos para añadirle a estos, tiempo de estudio durante varias noches por semana. Yo le recomiendo estudiar los idiomas, pero mientras tanto, he organizado este libro asumiendo que usted no lo ha hecho. Debo añadir el hecho de que la mayoría de cursos de idiomas no incluyen los términos usados en la construcción. Recomiendo enormemente que usted trate de comunicarse en el otro idioma la mayor cantidad del tiempo posible. Esto le permitirá mejorar en la construcción de frases y de oraciones y grabar sus ideas en la mente de sus compañeros de trabajo. Hable con la persona que habla en otro idioma acerca del almuerzo, de deportes y de otros temas. (Usted podrá hacer un nuevo amigo.)

Este libro consta de dos secciones: En la primera, las palabras se han agrupado en la forma como estas se utilizan en la vida real. Los hispanohablantes así como los anglohablantes, de forma similar, encuentran palabras al referirse a sus usos, a sistemas y a agrupaciones naturales. Estos capítulos reflejan la agrupación de los trabajos de construcción. Cada sección contiene la lista de las palabras necesarias y/ o de las frases para las personas que trabajan en dicha especialidad. Dentro de cada sección, las palabras se enumeran según su uso, tal como: palabras que se refieren a acciones (verbos), descriptores (adjetivos) y así sucesivamente. Aún dentro de las listas, he intentado enumerar y aparear las palabras que tienen relación entre sí. Luego, al referirse a los trabajos de mampostería, encontrará los tipos de ladrillo enumerados conjuntamente con otros materiales básicos. Esta agrupación combinada con las ilustraciones, debe lograr que la búsqueda de una palabra dada se convierta en un proceso intuitivo, de ser posible.

En la segunda sección las palabras se enumeran alfabéticamente para que usted pueda encontrar una palabra conocida, de forma inmediata. Existen dos listas: todos los términos en inglés enumerados alfabéticamente y los términos en español alfabéticamente.

En adición al cubrimiento de la construcción, este libro además trata otros temas necesarios, tales como el empleo. Esta es un área en la cual las personas tanto de habla hispana como inglesa interactúan inicialmente y es muy importante para las personas que trabajan actualmente (o que desean hacerlo) dentro de la industria de la construcción. Usted encontrará material sobre el negocio de construcción, mantenimiento, materiales, equipos, factores de conversión y medidas. He añadido una sección sobre el trabajo de demolición, que con frecuencia es el primer trabajo de construcción para los hispanohablantes.

Sospecho que existen términos importantes que he pasado inadvertidamente por alto dentro de este libro, si se tiene en cuenta que la terminología de construcción en cualquiera de los dos idiomas puede llenar varios libros. Por lo general, he tomado en su defecto los términos mexicanos cuando existen diferencias, más sin embargo, he tratado de incluir alternativas. Por lo tanto, me tocó hacer escogencias y permanezco abierto a las sugerencias de los lectores con relación a alternativas, deletreo y demás. Estaré actualizando de forma continua este libro e intentaré incluir el material que me sugieran mis lectores y mantenerme al día con los desarrollos de la industria de la construcción.

Paul Rosenberg

How To Use This Book

This book contains two sections. In the main section of this book (chapters three through seventeen), the words are grouped as they are used in real life. Spanish- and English-speakers alike find words by referring to their uses, systems and natural groupings. If you look at page 5-6, you will see that the section on carpentry begins with types of wood. As you proceed through the list (pages 5-7 and 5-8), you will notice that it covers other types of supports and building materials, then on to things that are made of these materials. The listings follow a natural progression.

The chapters reflect the breakdown of construction work. Each section should contain a list of necessary words and/or phrases for the people who work in that specialty. Within each section, the words are listed according to their use, such as words that refer to actions (verbs), to descriptions (adjectives) and so on. Even within lists, I attempted to list each word alongside words that have a relation to it. So, when looking at words related to masonry, you will find types of bricks listed together, with other basic materials listed next to the bricks. This grouping, combined with the illustrations, should make finding any given word as instinctual a process as is possible.

In the second main section of this book (Chapters 20 through 23), words are listed alphabetically, so that you can find a word you already know immediately. There are two lists: All English terms listed alphabetically, and all Spanish terms listed alphabetically.

In addition to all this, there are a few specialty chapters:

Chapter One, which covers employment and money.

Chapter Two, which covers the rules of pronunciation for English-Spanish and Spanish-English.

Chapter Eighteen, which covers conversion factors and other technical values in both Spanish and English.

Chapter Nineteen, which covers the characteristics of materials in both Spanish and English.

Cómo Utilizar Este Libro

Este libro contiene dos secciones. En el principal sección de este libro (capítulos 3 a 17), se agrupan las palabras mientras que se utilizan en vida verdadera. Los españoles y los Inglé's-locutores encuentran igualmente palabras refiriendo a sus aplicaciones, sistemas y agrupaciones naturales. Si usted mira la página 5-6, usted verá que la sección en la carpintería comienza con los tipos de madera. Mientras que usted procede a través de la lista (páginas 5-7 y 5-8), usted notará que cubre otros tipos de ayudas y de materiales de construcción, después encendido a las cosas que se hacen de estos materiales. Los listados siguen una progresión natural.

Los capítulos reflejan la interrupción del trabajo de construcción. Cada sección debe contener una lista de palabras y/o de frases necesarias para la gente que trabaja en esa especialidad. Dentro de cada sección, las palabras se enumeran según su uso, tal como palabras que refieran a las acciones (verbos), a las descripciones (adjetivos) etcétera. Incluso dentro de listas, procuré enumerar y redactar junto a las palabras que tienen una relación a él. Así pues, al mirar las palabras relacionadas con la albañilería, usted encontrará tipos de ladrillos enumerados juntos, con otras materias primas enumeradas al lado de los ladrillos. El este agrupar, combinado con las ilustraciones, debe hacer encontrar cualquier palabra dada tan instinctual un proceso como es posible.

En la segunda principal sección de este libro (capítulos 20 a 23), las palabras se enumeran alfabéticamente, de modo que usted pueda encontrar una palabra que usted sabe ya inmediatamente. Hay dos listas: Todos los términos ingleses enumeraron alfabéticamente, y todos los términos españoles enumerados alfabéticamente.

Además de todo el éste, allí es algunos capítulos de la especialidad:

Capítulo uno, que cubre el empleo y el dinero.

Capítulo dos, que cubre las reglas de la pronunciación para inglés-y Español-Inglés.

Capítulo dieciocho, que cubre factores de la conversión y otros valores técnicos en español e inglés.

Capítulo diecinueve, que cubre las características de materiales en español e inglés.

A note to our customers

We have manufactured this book to the highest quality standards possible. The cover is made of a flexible, durable and water-resistant material able to withstand the toughest on-the-job conditions. We also utilize the Otabind process which allows this book to lay flatter than traditional paperback books that tend to snap shut while in use.

Nota para nuestros clientes:

Hemos fabricado este libro bajo los más altos estándares de calidad posibles. La cubierta está fabricada con un material flexible, durable y resistente al agua, capaz de resistir las condiciones más duras en el trabajo. También utilizamos el proceso Otabind que permite que este libro permanezca abierto de manera más plana que los libros tradicionales de edición rústica, los cuales tienden a cerrarse mientras se utilizan.

CONTENTS

CHAPTER 4/CAPITULO 4
Site Work. **4-1**
Trabajo de Sitio **4-1**

CHAPTER 7/CAPITULO 7

Demolition. 7-1
Demolición 7-1

CHAPTER 8/CAPITULO 8

Welding 8-1
Soldadura 8-1

CHAPTER 9/CAPITULO 9
Finishing **9-1**
Acabado **9-1**

CHAPTER 10/CAPITULO 10
Plumbing **10-1**
Plomería **10-1**

CHAPTER 11/CAPITULO 11
HVAC and Mechanical 11-1
HVAC y Mecanico 11-1

CHAPTER 12/CAPITULO 12
Electrical 12-1
Electricista 12-1

CHAPTER 13/CAPITULO 13

Safety **13-1**
Seguridad **13-1**

CHAPTER 14/CAPITULO 14

Tools and Equipment **14-1**
Herramientas y Equipo **14-1**

CHAPTER 15/CAPITULO 15
The Construction Business **15-1**
El Negocio de Construcción **15-1**

CHAPTER 16/CAPITULO 16
Housekeeping **16-1**
Que Haceres. **16-1**

CHAPTER 17/CAPITULO 17
Basic Words and Phrases **17-1**
Palabras y Frases Básicas **17-1**

CHAPTER 18/CAPITULO 18
Measurement and Conversion Factors **18-1**
Unidades de Medida y Factor de Conversiones **18-1**

CHAPTER 19/CAPITULO 19
Materials and Tools **19-1**
Materiales y Herramientas........ **19-1**

CHAPTER 20/CAPITULO 20
English-Spanish Glossary
(Alphabetical Listing). **20-1**

CHAPTER 21/CAPITULO 21
Español-Inglés Glosario
(Listado Alfabético) **21-1**

CHAPTER 1/CAPITULO 1
Employment and Money
Empleo y Dinero

EMPLOYMENT — EMPLEO		
English	*Español*	
To hire, employ	Contratar, emplear	
Job	Trabajo	
Job site	Lugar de la obra	
Salary	Salario	
Union	Unión, sindicato	
Employer, boss	Jefe	Foreman Capataz
Foreman	Capataz	
Name	Nombre	
Office	Oficina	
Secretary	Secretario, secretaria	Office Oficina
Address	Dirección	
Phone number	Número de teléfono	
Manager	Gerente	

EMPLOYMENT (cont.) — EMPLEO (cont.)

English	Español	
Superintendent	Superintendente	
Interpreter	Intérprete	
Today	Hoy	
Tomorrow	Mañana	
Tools	Herramientas	Tools Herramientas
Training	Entrenamiento	
Safety policy	Póliza de seguridad	
To read	Leer	
To write	Escribir	
Family	Familia	Family Familia
Man	Hombre	
Woman	Señorita, señora	
Children	Niños	

PAYMENT — PAGO		
English	**Español**	
Money	Dinero	
Payment	Pago	
Agreement	Acuerdo	Money/ Cash Dinero/ Efectivo
Contract	Contratar	
Dollars	Dólares	
Cash	Efectivo	
Change	Cambio	Dollars Dolares
Check	Cheque	
Withholding	Retener	
Taxes	Impuestos	
Federal taxes	Impuestos federales	Change Cambio
State taxes	Impuestos estatales	
Local taxes	Impuestos locales	
Medicare	Medicare	Check Cheque

PAYMENT (cont.) — PAGO (cont.)

English	Español	
Social security number	Número de Seguro Social	Social security number / Número de Seguro Social
Insurance	Seguro	
Health insurance	Seguro de salud	
Bank	Banco	Bank / Banco
Bank account	Cuenta de banco	
Pension	Pensión	
Payday	Día de paga	

LEGAL — LEGAL

English	Español	
Green card	Tarjeta de residencia	Green card / Tarjeta de residencia
Legal	Legal	
Illegal	Ilegal	
Visa	Visa	
Passport	Pasaporte	Passport / Pasaporte
Citizen	Ciudadano	
Citizenship	Ciudadanía	

BANKING AND BUSINESS — CUENTA BANCARIA Y NEGOCIOS

English	Español
Signature	Firma
Date	Fecha
Amount	Cantidad, suma
Name	Nombre
Payee	Beneficiario
Bank name	Nombre de banco
Account number	Número de cuenta
Check number	Número de cheque

Check — Cheque

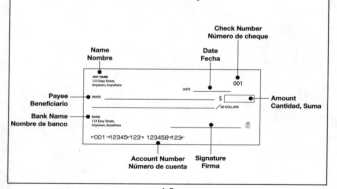

BANKING AND BUSINESS *(cont.)* — CUENTA BANCARIA Y NEGOCIOS *(cont.)*

English	Español	
Deposit	Depósito	
Withdrawal	Retirada	Beginning Balance $1,700.87 Deposits $3,658.62 Withdrawals $3,478.95 Ending Balance $1880.54 Balance Balance
Promise	Promesa	
Owe	Deber	
Loan	Préstamo	
Balance	Balance	
Buy	Comprar	
Sell	Vender	
Store	Tienda	SALE Store Tienda
To rent	Rentar	

PHRASES — FRASES

English	Español
Do you speak English?	¿Habla inglés?
What is your name?	¿Cómo se llama?
My name is ___.	Mi nombre es ___.
Pleased to meet you.	Mucho gusto.
What is your phone number?	¿Cuál es tu número de teléfono?
I need you to fill out this form.	Necesito que complete este formulario.
I need to see the identification you listed on the form.	Necestio ver la identificación que indicó en el formulario.
Without ID, I can't hire you.	Sin la identificación adecuada, no puedo emplearlo.
Do you have your own tools?	¿Tiene sus propias herramientas de mano?
Your pay will be ___ per hour.	Se la va a pegar ___ por hora.
… less tax withholdings.	… menos descuentos por impuestos.

PHRASES *(cont.)* — FRASES *(cont.)*

English	Español
Can you work tomorrow?	¿Puede trabajar mañana?
Can you drive a car?	¿Sabe conducir?
Do you have a driver's license?	¿Tiene licencia de conducir?
Come with me.	Venga conmigo.
Are you thirsty?	¿Tiene sed?
Are you hungry?	¿Tiene hambre?
Wear this for your protection.	Use esto para su protección.
Do you speak English?	¿Hablas Inglés?
Do you speak Spanish?	¿Hablas Español?
Work safely.	Trabaje con cuidado.
Can someone interpret?	¿Puede alguien interpretar?

CHAPTER 2/CAPITULO 2
Pronunciation
Pronunciación

PRONUNCIACIÓN EN INGLÉS

Acentos:

Encontrará acentos sólo en las palabras que pudiera tener dudas.

Vocales:

a es una vocal abierta entre la *a* y la *e*. La cual no se encuentra en español.

a̲ es una vocal cerrada y resuena en la parte posterior de la cavidad bucal. La cual no existe en español.

e es una vocal cerrada. Despues de la *o* solo se usa para cerrar el diptongo (doble vocales).

o es una vocal abierta casi como para decir *a*.

iy es una *i* larga.

Otros simbolos usados:

dy es un sonido fuerte de la y o *ll*.

s su sonido es similar al de una abeja.

th como la z en castellano (thin) o **th** como la d en la palabra (then).

PRONOUNCING SPANISH

Vowels:

A is pronounced	AH as in far
E is pronounced	EH as in mend
I is pronounced	EE as in feet
O is pronounced	OH as in only
U is pronounced	OO as in pool

Diphthongs (double vowels):

io is pronounced	EEOH as in
ie is pronounced	yeh as in
iu is pronounced	WEE as in
ua is pronounced	WAH as in
ue is pronounced	WEH as in

Consonants:

h is always silent

j is pronounced as an *h*

ll is pronounced *y* as in yell

ñ is pronounced *nya*, as in canyon

ny is pronounced *nya* as in onion

d is voiced

t and **p** are soft

r is soft

rr is rolled

ch is always as in Church

Plurals:

el changes to **los**

la changes to **las**

s or **es** are added to the end of the word.

CHAPTER 3/CAPITULO 3
General Construction
Construcción General

CONSTRUCTION DRAWINGS — DIBUJOS DE CONSTRUCCIÓN		
English	**Español**	
Drawings	Dibujos	Drawings Dibujos
Architectural plans	Planos arquitectónicos	
Site plans	Planos de sitio	
Structural plans	Planos estructurales	
Symbols	Símbolos	
Schedules	Horarios	
Section	Sección	
Shop drawings	Dibujos de taller	
Front view	Vista de frente	
End view	Vista de costado	Symbols Símbolos
Overhead view	Vista de arriba	
Section	Sección	
Details	Detalles	
Dimensions	Dimensiones	
Directions (North…)	Direcciones (Norte…)	
North	Norte	
South	Sur	
East	Este	Directions (North…) Direcciones (Norte…)
West	Oeste	

3-1

OTHER DOCUMENTS — DOCUMENTOS OTROS

English	Español	
Schedule	Horario	
Construction schedule	Cronograma de construcción	
Estimate	Estimación	
Bid	Oferta	
Contract	Contrato	
Notes	Notas	
Notice	Aviso	
Permit	Permiso	Contract Contrato
Lien	Derecho de retención	
Bond	Bono	
Performance bond	Bono de ejecución	
Report	Reportaje	
Job report	Reportaje de trabajo	
Daily report	Reporte del día	
Notice to proceed	Aviso de proceder	
Change order	Orden de cambiar	
Safety policy	Póliza de seguridad	Change order Orden de cambiar

PEOPLE — PERSONAS

English	Español	
Architect	Arquitecto	Architect / Arquitecto
Engineer	Ingeniero	
Owner	Dueño	
Contractor	Contratista	
Inspector	Inspector	
Building inspector	Inspector de obras	
Estimator	Estimador	Contractor / Contratista
Supervisor	Supervisor, supervisora	
Manager	Gerente	
Foreman	Capataz	
Journeyman	Oficial	
Apprentice	Aprendiz	Foreman / Capataz
Carpenter	Carpintero	
Worker	Trabajador	
Technician	Técnico	
Helper	Ayudante	
Manufacturer	Fabricante	Worker / Trabajador

LOCATIONS — LOCACIONES

English	Español	
Job site	Lugar de la obra	
Property	Propiedad	
Property line	Linea de propiedad	
Tool lockup	Cuarto de herramientas llavado	Lockers Gavetas
Lockers	Gavetas	
Street	Calle	
Road	Camino	Slab Losa
Alley	Callejón	
Walkway	Acera	
Driveway	Vía de acceso	Parking lot Estacionamiento
Slab	Losa	
Parking lot	Estacionamiento	
Sewer	Alcantarilla	
Footing	Zapata	Footing Zapata
Foundation walls	Muros de fundación	
Ground elevation	Rasante	
Walls below grade	Muros por debajo del nivel de terreno	Ground elevation Rasante

English	Español	
Wall	Pared	Wall Pared
Exterior wall	Muro exterior	
Entry	Entrada	
Exit	Salida	
Basement	Sótano	
Boiler room	Cuarto de calderas	Stairs Escaleras
Roof	Techo	
Stairs	Escaleras	
Floor	Piso	
Crawl space	Espacio angosto	Roof Techo
Room	Cuarto, sala	
Bathroom	Cuarto de baño	
Window	Ventana	Window Vertana
Corridor	Pasillo	
Door	Puerta	
Doorway	Bano	
Ceiling	Cielo	Door Puerta
Kitchen	Cocina	

LOCATIONS (cont.) — LOCACIONES (cont.)

English	Español	
Hallway	Vestíbulo	
Bedroom	Habitación, dormitorio	
Vestibule	Vestíbulo	
Mezzanine	Entresuelo	
Elevator	Elevador	
Walk-in cooler	Frigorífico	
Moving walkway	Caminos moviles	
Penthouse	Sobradillo	Show window / Vitrina
Aisle	Pasillo	
Atrium	Atrio	
Attic	Ático	
Canopy	Toldo	
Show window	Vitrina	
Façade	Alzado	
Kiosk	Quiosco	Gymnasium / Gimnasio
Classroom	Sala de clase	
Gymnasium	Gimnasio	

LOCATIONS (cont.) — LOCACIONES (cont.)

English	Español	
Storage room	Cuarto de almacenamiento	
Dwelling	Vivienda	House / Casa
Dwelling unit	Unidad de vivienda	
Occupancy	Destino, tenencia	
House	Casa	
Apartment building	Edificio de departamentos	Apartment Building / Edificio de departamentos
Condominium	Condominio residencial	
High-rise building	Edificio de gran altura	
Store	Tienda	High-rise building / Edificio de gran altura
Mall	Centro commercial	
Dormitory	Residencias para estudiantes	
Reformatory	Reformatario	
Warehouse	Deposito, bodega	
Swimming pool	Piscina (de natación), alberca	Swimming pool / Piscina (de natación), alberca

ACTIONS — ACCIONES

English	Español
Communicate	Comunicar
Coordinate	Coordinar
Add	Agregar
Apply	Aplicar
Fasten	Atar
Check	Comprobar
Warn	Advertir
Help	Ayudar
Train, teach	Entrenar
Require	Requerir
Turn off	Apagar
Turn on	Prender
Draw	Dibujar
Mark	Marcar
Write	Escribir
Document, record	Documentar
Maintenance	Mantenimiento
Label	Etiquetar
Clean	Limpiar
Cover	Cubrir
Anchor, affix	Anclar
Dial (phone)	Marcar (teléfono)
Examine	Examinar

ACTIONS (cont.) — ACCIONES (cont.)

English	Español
Verify	Verificar
Ventilate	Ventilar
Lift	Levantar
Roll	Rodar
Put on	Poner
Locate	Localizar
Modify	Modificar
Test	Probar
Tie off	Atarse
Point	Señalar
Wear	Usar (ropa)
Divide	Dividir
Extend	Extender
Recover	Recuperar
Verify	Verificar

DESCRIPTIONS — DESCRIPCIONES

English	Español
Competent	Competente
Incompetent	Incompetente
Early	Temprano
Late	Atrasado
Missing	Perdido, faltar
Motivated	Motivado
On time	A tiempo
Quickly	Rápido
Ready	Listo
Safely	Con seguridad
Slowly	Lento
Approved	Aprobado
Rejected	Rechazarado
Interior	Interior
Exterior	Exterior
Underground	Subterráneo
Located	Localizado
Parallel	Paralelo
Perpendicular	Perpendicular
Heavy	Pesado
Loose	Flojo
Loud	Ruidosa
Sanitary	Sanitario

SIGNS AND WARNINGS — SIGNOS Y AVISOS

English	Español
Sharp	Filoso
Teamwork	Trabajo en equipo
Tight	Apretado
Authorized personnel only	Personal autorizado solamente
Asbestos	Asbestos
Broken glass	Cristal roto
Danger	Peligro
Hazardous	Pelligrosa(o)
Keep out	Mantengase fuera
Overhead work	Trabajo de arriba
Pedestrian walkway	Calzada
Protruding nails	Clavos salientes
Hard hat area	Área de cascos
Explosives	Explosivos
Evacuate	Evacuar
Lockout/tagout	Cierre
No smoking area	Zona de no fumar
Warning	Aviso
Live wires	Alambres vivos
Paint thinners	Diluyentes de pintura
Gasoline	Gasolina

PHRASES — FRASES

English	Español
Everything is ready.	Todo está listo.
Get it done quickly and safely.	Cumplalo rápidamente y con seguridad.
Wear this for your protection.	Use este para su protección.
Check the plans.	Compruebe revise los planos.
Figure it out.	Dedúzcalo.
Finish up.	Terminar.
Get it done.	Cumplelo. Hazlo.
Hurry up.	Apurarse.
This is your responsibility.	Esta es su responsibilidad.
I need you here early to open the gate.	Te necesito aquí temprano para abrir la puerta.

PHRASES (cont.) — FRASES (cont.)

English	Español
Can you work tomorrow?	¿Puede trabajar mañana?
Can you drive a car?	¿Sabe conducir?
Do you have a driver's license?	¿Tiene licencia de conducir?
Come with me.	Venga conmigo.
Are you thirsty?	¿Tiene sed?
Are you hungry?	¿Tiene hambre?
The plans and specs don't agree.	Los planos y las especificaciones no concuerdan.
Check your contract.	Compruebe revise su contrato.

CONSTRUCTION COMPONENTS

Ceiling joist/
Vigueta de cielo

Rafter/
Cabrio, cabio

Underlayment/
Suelo subyacente

Drywall/
Pirca

Wood studs/
Montanes, pies
derechos
de madera

Soffit/
Sofito

Wall brace/
Apoyo de pared

Plate/
Placa

Foundation wall/
Muro de fundación

Girder/
Viga principal

Footing/
Zapata

Subfloor/
Subsuelo

Joist/
Vigueta

COMPONENTES DE CONSTRUCCIÓN

Fascia/
Omposta

Header/
Cabezal

Roof sheathing/
Entablado de techo

Roof/
Techo

Gutter/
Gotera

Door jamb/
Jamba de
puerta

Window casing/
Chambrana
de ventana

Siding/
Forrado,
revestimiento

Sheathing/
Entablado

Drain/
Desagüe

Waterproofing/
Impermeable

Bridging/
Arriostramiento

3-15

NOTES — NOTAS

CHAPTER 4/CAPITULO 4
Site Work
Trabajo De Sitio

EXCAVATION AND EARTH WORK — EXCAVACION Y TRABAJO DE TEIRRA		
English	*Español*	
Property	Propiedad	
Property boundary line	Linea de limite de la propiedad	Plot Trazar
Boundary	Límite	
Elevation	Elevación	
Plot	Trazar	
Lot	Terreno, lote	
Survey	Deslindar	
Bracing	Riostras	
Earth work	Terrapien	
Cave-in	Colapso de agujero	
Egress	Salida	Flag stake Estaca de bandera
Shoring	Puntales	
Swing	Oscilación	
Flag stake	Estaca de bandera	
Wood stake	Estaca de Madera	
String line	Linea de hilo	
Grid lines	Lineas de rejilla	Power lines Líneas de energía
Power lines	Líneas de energía	

EXCAVATION AND EARTH WORK (cont.) — EXCAVACION Y TRABAJO DE TEIRRA (cont.)

English	Español	
Underground lines	Líneas subterráneas	
Pipeline	Tubería	
Manhole	Pozo de confluencia, boca de inspección, boca de accesso, pozo de entrada	Pipeline Tuberia
Sanitary sewer	Sistema de alcantarilla sanitario	
Sewage	Aguas negras	
Storm drain	Drenaje para tormentas	
Drain, drainage	Desagüe	Storm Drain Drenaje para tormentas
Public utilities	Servicios públicos	
Waterline	Pipa de agua	
Septic tank	Fosa septica	
Spray paint	Pintura rociada	
Paint mark	Pintura de marcar	
Theodolite	Teodolito	Septic tank Fosa septica
Blue stakes	Estacas azules	
Caution tape	Cinta de precaución	
Dust control	Eliminación de polvo	
Hazardous communications	Comunicaciones peligrosas	CAUTION
Soil type	Tipo de suelo	Caution tape Cinta de precaución
Stability	Estabilidad	

EXCAVATION AND EARTH WORK (cont.) — EXCAVACION Y TRABAJO DE TEIRRA (cont.)

English	Español	
Embankment	Terraplén	
Incline	Declive, inclinación	 Incline Declive, inclinación
Shored construction	Construcción apuntalada	
Unshored construction	Construcción no apuntalada	
Backup alarm	Alarma de refuerzo	
Traffic awareness	Conocimiento de tráfico	
Hydrant	Boca de riego	 Hydrant Boca de riego
Templates	Plantillas	

PEOPLE — PERSONAS

English	Español	
Flagger	Persona que da señales	
Surveyor	Topógrafo	 Surveyor Topógrafo
Landscaper	Paisajista	
Geotechnical engineer	Ingeniero de geotécnico	
Soil engineer	Ingeniero de Suelos	

SURFACE FEATURES — CARACTERÍSTICAS DE LA SUPERFICIE

English	Español	
Soil	Suelo	
Bedrock	Roca de fondo	
Boulder	Pedron	Rock
Slope	Cuesta, Inclinación	Roca, piedra
Rock	Roca, piedra	
Stone	Piedra	
Sandstone	Arenisca	
Limestone	Caliza	
Water table	Tabla de agua	Dirt, earth
Clay	Arcilla	Tierra
Dirt, earth	Tierra	
Sand	Arena	
Clod	Terrón	
Dust	Polvo	
Topsoil	Suelo Vegetal	Trees
Trees	Árboles	Arboles
Limbs	Miembros	
Weeds	Malas hierbas	Slab
		Losa

4-4

SURFACE FEATURES (cont.) — CARACTERÍSTICAS DE LA SUPERFICIE (cont.)

English	Español	
Quicksand	Arena movediza	
Slab	Losa	
Steam	Vapor	
Fire	Fuego	Shrubs / Arbustos
Ice	Hielo	
Thawing	Descongelación	
Shrubs	Arbustos	
Soil	Suelo	
Sod	Césped	
Surface water	Agua superficial	Slope / Pendiente, talud
Trench/ditch	Zanja	
Slope	Pendiente, talúd	
Slump	Asentamiento	
Unstable ground	Terreno inestable	
Drainage	Drenaje	
Erosion	Erosión	
Veins	Venas	Rubbish / Desperdicios
Rubbish	Desperdicios	

4-5

SURFACE FEATURES *(cont.)* — CARACTERÍSTICAS DE LA SUPERFICIE *(cont.)*

English	Español	
Rubble	Escombro	Rubble / Escombro
Moisture	Humedad	
Runoff	Agua de desagüe	
Bench mark	Punto de referencia	
Fill, backfill	Relleno	
Unbalanced fill	Relleno sin consolidar	Grade / Grado
Grade	Grado	
Tunnel	Tunel	
Culvert	Alcantarilla	
Curve	Curva	
Depth	Profundidad	Culvert / Alcantarilla
Distance	Distancia	
Drill	Taladro	
Utility lines	Lineas de servicios publicos	
Fence	Cerca	
Hole	Agujero	Fence / Cerca
Pit	Hoyo, pozo	
Pipe	Tubo	

TOOLS AND EQUIPMENT — HERRAMIENTAS Y EQUIPO

English	Español	
Backhoe	Retroexcavadora	
Blade	Hoja	
Bucket	Cubo	**Backhoe** Retroexcavadora
Chain	Cadena	
Dump truck	Camión de descarga	
Truck bed	Caja de camioneta	
Coolant	Liquido refrigerante	
Diesel	Diesel	**Dump truck** Camión de descarga
Gasoline	Gasolina	
Generator	Generador	
Machine	Máquina	
Grader	Máquina niveladora	
Gravel	Grava	
Loader	Cargador	**Generator** Generador
Ladder	Escalera	
Mud	Lodo	
Pile	Pila	
Ripper	Destripador	
Roller	Rodillo	**Ladder** Escalera
Ramp	Rampa	

TOOLS AND EQUIPMENT *(cont.)* —
HERRAMIENTAS Y EQUIPO *(cont.)*

English	Español	
Shovel	Pala	
Tamper	Pisón	
Track hoe	Encarrilada	
Winch	Torno	Tripod Tripode
Water truck	Tanquero de agua	
Transit	Transito	
Tripod	Tripode	
Total station	Estación Total	
Hub	Eje	Topographic map Mapa Mapa topográfico
Topographic map	Mapa topográfico	
Walkie talkie	Transceptor portátil	
Lawnmower	Cortador de césped, cortacésped	
Fertilizer	Fertilizante, abono	Conveyor Teansportador de mecanico
Propane	Propano	
Matches	Fósforos	
Conveyor	Transportador de mecanico	
Dragline	Excavadora de arrastre	
Dredge	Draga	Walkie talkie Transceptor portátil

ACTIONS — ACCIONES

English	Español
Clear	Despejar
Cut	Cortar
Dig	Cavar
Expose	Exponer
Fill	Llenar
Bury	Enterrar
Drain	Drenar
Dump	Descargar
Grade	Nivelar
Haul	Acarrear
Drag	Halar
Scrape	Raspar
Shovel	Traspalar
Compact	Compactar
Cover	Cubrir
Dewater	Desaguar
Dig	Cavar
Drill	Agujerar

ACTIONS (cont.) — ACCIONES (cont.)

English	Español
Drilling	Perforación
Dump	Descargar
Rake	Rastillar
Decompose	Descomponerse
Stabilize	Estabilizar
Tamp	Apisonar
Tunnel	Cavar un túnel
Fill	Llenar
Move	Mover
Survey	Topografar
Transplant	Transplantar
Drive	Conducir
Lower	Bajar
Pull	Jalar
Push	Empujar
Release	Soltar
Replace	Reponer, reemplazar
Rescue	Salvar

ACTIONS (cont.) — ACCIONES (cont.)

English	Español
Bring	Traer
Stack	Apilar
Hold	Detener
Lay out	Hacer disposiciones
Mark	Marcar
Remove	Quitar
Sink	Hundirse
Spray	Rociar
Shoot (a line)	Tirar (una línea)
Spray	Rociar
Drive (a stake)	Golpear (una estaca)
Grease	Engrasar
Pile	Apilar
Put	Colocar
Pour	Echar
Pump	Bombear
Watch	Tener cuidado
Lubricate	Lubricar

DESCRIPTIONS — DESCRIPCIONES

English	Español
Dry	Árido
Dusty	Polvoriento
Exposed	Expuesto
Hard	Duro
Level	Nivel
Loose	Flojo
Muddy	Fangoso
Wet	Mojado
Subgrade	Subsuelo
Soft	Suave
Solid	Sólido
Behind	Detrás de
Stuck	Pegado
Broken	Roto, averiado
Damaged	Dañado
Dark	Oscuro
Dirty	Sucio
Extra	Sobra

DESCRIPTIONS *(cont.)* — DESCRIPCIONES *(cont.)*

English	Español
Traction	Tracción
On line	En línea
Fragile	Frágil
Front	Frente, delante
Tandem	Tándem
Height	Altura
Width	Anchura
Middle	Medio
Near	Cerca, proximo
One quarter	Un cuarto
Quickly	Pronto
Quiet	Tranquilo
Acre	Acre
Length	Longitud

PHRASES — FRASES

English	Español
This ground is hard and solid.	Esta tierra es dura y solida.
That section is cut to grade.	Este sección está cortada a nivel.
This ground is muddy.	Esta tierra esta lodosa.
Be sure to call before you dig.	Asegurese de llamar antes de que usted cave.
We need another egress ladder.	Necesitamos otra escalera de salida.
What will the width of this channel be?	¿Cual va a ser lo ancho de este canal?
Use bracing if there could be a cave-in.	Utilice riostras si hay cualquier posibilidad de colapso.
Snap a line.	Echar una raya.
Will you deliver and lay the sod?	¿Puede entregar y poner el césped?
Hold the grade stick plumb.	Detenga el palo de grado plomado.
Change your radio battery.	Cambie la pila de su radio.

CHAPTER 5/CAPITULO 5
Structural and Carpentry
Estructura y Carpinteria

STEEL CONSTRUCTION — CONSTRUCCIÓN DE ACERO		
English	*Español*	
Structure	Estructura	Girder / Viga principal Beam / Viga Column / Columna **Structural Members** **Miembro de estructura**
Beam	Viga	
Girder	Viga principal	
Column	Columna principal	
Caissons	Cajones de aire comprimido	
Joist	Vigueta	
Joist bridging	Arriostramiento de vigueta	
Truss	Armadura de cubierto	**Joist** **Vigueta**
I beam	Viga doble T	
Grade beam	Viga de fundación	
Structural steel	Acero estructural	
Span	Abarbetado	**Truss** **Armadura de cubierto**
King post	Poste principal	
Queen post	Columna	
Plate girder	Viga de alma llena	
Anchor bolt	Perno de anclaje	**I beam** **Viga doble T**

English	Español	
Angle iron	Ángulo de hierro	
Banding	Banda de ligadura	
Bolt	Perno	
Bracing	Apoyo	**Bolt** / **Perno**
Live load	Cargas vivas	
Combination load	Combinación de cargas	
Loaded area	Área cargada	
Earthquake load	Carga sísmica	**Floor** / **Piso**
Occupant load	Número de ocupantes	
Connection	Conexión	
Decking	Tablero de lamina	
Floor deck	Plataforma	
Floor	Piso	
Span	Abarbetado	**Floor Deck** / **Platform**
Degrees (angle)	Grados (de ángulo)	
Right angle	Ángulo recto	
Cantilever	Voladizo	
Roof	Azotea	
Anchorage point	Punto de anclaje	90°
Beater	Martillito	**Right angle** / **Ángulo recto**
Embed plate	Placa de anclaje	

STEEL CONSTRUCTION (cont.) — CONSTRUCCIÓN DE ACERO (cont.)

English	Español	
Shackle	Grillete	
Camber	Comba	
Retractable reel	Carrete retractable	135°
Sharp edge	Borde agudo	Degrees (angle) Grados (de ángulo)
Burr	Rebaba	
Degrees (angle)	Grados (de ángulo)	
Nut	Tuerca	
Roof	Azotea	
Hammer drill	Rotamarillo	
Impact wrench	Arranque de impacto	Roof Techo
Nut	Tuerca	
Lanyard	Acollador	
Pinch point	Punto de pellizco	
Temporary power	Energia temporaria	
Scaffolding	Andamios	Hard hat Casco
Soap stone	Jaboncillo	
Hard hat	Casco	
Extension ladder	Escalera de extensión	
Square	Escuadra	
Washer	Rondana	
C clamp	Grapa en C	C clamp Grapa en C

STEEL CONSTRUCTION *(cont.)* — CONSTRUCCIÓN DE ACERO *(cont.)*

English	Español	
Bags	Bolsas	Cords / Cuerdas
Cords	Cuerdas	
Fall protection	Protección de caída	
Guard rail	Barandal	
Hard hat	Casco	
Torch	Antorcha	
Square	Escuadra	
Wrench	Llave inglesa	
Torque wrench	Llave dinamométrica	
Spud wrench	Llave de cola	Torch / Antorcha
Rivet	Remanche	
Die/chaser nut	Dado fraccionario	
Tag	Etiqueta	
Choker	Cable ahogador	
Sleever bar	Palanca de cola	
Sling	Eslinga	Choker / Cable ahogador
Bull pin	Perno de Alineación	
Tag line	Línea de guía	
Come-along	Trinquete a polea	

STEEL CONSTRUCTION (cont.) — CONSTRUCCIÓN DE ACERO (cont.)

English	Español	
Crane	Grúa	
Crane operator	Operador de grúa	
Crane hand signals	Señales de mano de grúa	
Lay out	Croquis	Crane Grúa
Sheer wall	Muro cortante	
Shaft	Recinto	
Shell	Cáscara, cubierta	
Spandrel	Jacena exterior, timpano	
Tension	Tensión	Spandrel Jacena exterior, timpano
Lintel	Dintel	
		Lintel Dintel

CARPENTRY — CARPINTERÍA

English	Español	
Lumber	Madera de construcción	
Graded lumber	Madera elaborada	
Pine	Pino	
Oak	Roble	Lumber Madera de construcción
Redwood	Madera de secoya	
Cross grain	Fibra transversal	
Treated lumber	Madera de construcción tratada	
Wood studs	Montajes, Columnas de madera	
Steel studs	Montajes, pies derechos de acero	Stud track Carril de montante
Stud track	Carril montante	
Frame	Marco, estructura	
Framing	Formando a la construcción	
Framework	Armazón	
Furring, furred	Enrasado	
Board	Panel	
Planking	Entablonado	Steel studs Montajes, pies derechos de acero
Rafter	Cabio	
Joist	Vigueta	
Double joist	Vigueta doble	
Load-bearing joist	Viga de carga	Rafter Cabio

CARPENTRY (cont.) — CARPINTERÍA (cont.)

English	Español	
Joist hanger	Soporte de vigueta	
Pole, post	Poste	
Header	Cabezal	
Top plate	Placa superior	
Truss	Armadura de cubierto	
Rafter	Cabrio, cabio	Pole, post Poste
Cripple stud	Mantantes cojos	
Wall brace	Apoyo de pared	
Toe nail	Clavo oblicuo	Truss Armmadura de cubierto
Plywood	Chapeado	
Ply	Capa	
Sheathing	Entablado	
Sheet	Pilego, chapa, plancha	Plywood Chapeado
Sheeting	Revestimiento, laminado	
Subfloor	Subsuelo	
Decking	Tablero	
Fascia	Omposta	
Particle board	Tablero prensado	
Stud wall	Muro con montanes	
Bearing wall	Banda de ligadura	
Parapet wall	Pared de parapeto	
Top plate	Placa	Stud wall Muro con montanes
Blocking	Travesaño, obstruyendo	

CARPENTRY (cont.) — CARPINTERÍA (cont.)		
English	**Español**	
Fire stop	Tope antifuego	
Bracing	Apoyo	
Bridging	Arriostramiento	
Post	Poste	
Crown	Vértice	Post Poste
Stair stringer	Zanca de escalera	
Partition	Partición	
Door	Puerta	
Door frame	Marco de puerta	
Door jamb	Jamba de puerta	
Doorway	Portal	
Door sill	Umbral	
Draft stop	Cierro de tiro	Door jamb Jamba de puerta
Window	Ventana	
Foundation	Fundación	
Gable	Astial	
Rough opening	Abertura bosqueja	
Rough sill	Umbral bosquejo	Window Ventana
Riser (stair)	Contrahuella	
Landing (stair)	Descanso de escaleras	
Jamb	Jamba	
Sill	Umbral	Riser (stair) Contrahuella

CARPENTRY (cont.) — CARPINTERÍA (cont.)

English	Español	
Rabbet	Muesca, ranura	
Undercut	Resquicio	
Sky light	Celocia	
Soffit	Sofito	
Window	Ventana	Sky light / Lucernario
Window frame	Marco de ventana	
Mortise	Ranura	
Mullion (door)	Larguero central	
Soleplate	Solera	
Bow	Arco	Pitch / Pendiente del techo
Rim beam	Viga del borde	
Pitch	Pendiente del techo	
Gusset	Cartabon	
Hat channel	Perfíl a sombrero	
C channel	Perfíl en C	
Ceiling suspension	Alambre de la suspensión del techo	C channel / Perfíl en C
Water level	Nivel de agua	
Acoustical panels	Paneles acústicos	
C clamp	Grapa en C	
Chalk box	Carrete entizado	C clamp / Grapa en C
Construction adhesive	Pegamento de construcción	

CARPENTRY (cont.) — CARPINTERÍA (cont.)

English	Español	
Mastic	Mástique	
Hammer	Martillo	Hammer Martillo
Hand saw	Sierra de mano	
Hand tools	Herramientas de mano	
Spike	Clavo especial para madera	
Spiked	Clavado	
Nail gun	Clavador neumático	
Nails	Clavos	Nails Clavos
Powder nailer	Pistola de cartuchos para fijación	
Nuts	Tuercas	
Saw horse	Banqueta de aserrado	
Scaffolding	Andamios	
Screws	Tornillos	Screws Tornillos
Screw driver	Desarmador	
Screw gun	Pistola descarmadora	
Knot	Nudo de Madera	
Pencil	Lápiz	
Mitre saw	Sierra de retroceso para ingletes	
Circular saw	Sierra circular de mano	Circular saw Sierra circular de mano
Worm-drive saw	Sierra circular con tornillo sinfin	

CARPENTRY (cont.) — CARPINTERÍA (cont.)

English	Español	
Circular saw blade	Disco	
Chop saw	Serrucho tajadero	
Radial arm saw	Serrucho guillotina	
Reciprocating saw	Sierra alternativa	
Table saw	Serrucho de mesa	Reciprocating saw Sierra alternativa
Saw guard	Protector de serrucho eléctrico	
Electric drill	Taladro eléctrico	
Screw gun	Pistola descarmadora	
Nail gun	Clavador neumático	
Tape measure	Cinta de medir	
Carpenter's apron	Mandil, delantal	Electric drill Taladro electrico
Carpenter's square	Escuadra de carpintero	
Framing square	Escuadra para formar	
Square	Escuadra	
Termite protection	Protección contra termes	
Swelling	Expansión	Square Escuadra
Templates	Plantillas	

ACTIONS — ACCIONES

English	Español
Anchor	Anclar
Bolt	Enparnar
Cut	Cortar
Fabricate	Fabricar
Fasten	Sujetar
Grind	Moler
Overlap	Traslapar, ensimar
Place	Colocar
Raise	Levantar
Mark	Marcar
Stand	Erguir
Tighten	Apretar
Unload	Descargar
Signal	Señalar con la mano
Rig	Aparejar
Rack	Gualdrapear
Plumb	Plomear
Lower	Bajar

ACTIONS (cont.) — ACCIONES (cont.)

English	Español
Connect	Conectar
Shim	Acuñar
Torque	Apretar (una tuerca)
Anchor	Anclar
Bolt	Empernar
Cut	Cortar
Fabricate	Fabricar
Fasten	Sujetar
Grind	Moler
Overlap	Traslapar, ensimar
Place	Colocar
Raise	Levantar
Mark	Marcar
Stand	Erguir
Tighten	Apretar
Unload	Descargar
Anchor	Anclar
Check	Comprobar, revisar

ACTIONS (cont.) — ACCIONES (cont.)

English	Español
Cut	Cortar
Drill	Taladrar
Fasten	Sujetar
Frame	Sujetar, formar
Hang (door)	Colgar (puerta)
Mark	Marcar
Measure	Medir
Nail	Clavar
Plug in	Enchufar
Raise	Alzar
Screw	Atornillar
Saw	Aserrar
Stack	Apilar
Tighten	Apretar
Unload	Descargar

DESCRIPTIONS — DESCRIPCIONES

English	Español
Straight	Derecho
Warped	Torcido
Square	Cuadrado
Level	A nivel
Plumb	Plomada
Loose	Flojo
Tight	Apretado
Set	Puesto
Solid	Sólido
Weak	Débil
Cambered	Combado
Secured	Asegurado
Securely	Seguramente
Rough	Áspero
Shiny	Brilliante
Smooth	Liso
Stuck	Pegado
Tight	Aprieto
Even	Uniforme
Uneven	Desigual
Quickly	Rápidamente
Slowly	Lentamente
Fabricated	Fabricado
Degrees (angle)	Grados (de ángulo)

PHRASES — FRASES

English	Español
Is the ladder secured at the top?	¿Está la escalera fija arriba?
Can you rig and fly this joist?	¿Puede usted aparejar y volar esta Vigueta?
I need more three-inch bolts.	Necesito más tornillos de tres pulgadas de largo.
The column has a loose connection.	La columna tiene la conexión floja.
Do you know crane hand signals?	¿Sabe usted señales manuales de grúa?
This cord doesn't have a ground prong.	Este cable eléctrico no tiene la cuchilla a tierra.
These one by eights are all warped.	Todos estos uno por ochos están torcidos.
Never tamper with the guard on a Skill saw.	Nunca juegues con la guarda del cerrucho eléctrico.
Measure that stud again.	Mide ese montante otra vez.
Use screws instead of nails.	Utiliza tornillos en vez de clavos.
Double check.	Recomprobar revisar otra ves.
That is a damaged electrical cord.	Este es un cable eléctrico dañado.
Check the ground pin of the electrical plug.	Recomprobar revisa otra ves la cuchilla a tierra de clavija eléctrico.

CHAPTER 6/CAPITULO 6
Concrete and Masonry
Concreto y Albaneria

CONCRETE WORK — TRABAJO DE CONCRETO		
English	**Español**	
Concrete	Concreto	
Aggregates	Agregados	
Cement	Cemento	
Concrete mixing truck	Mezcladora de concreto sobre camión	Concrete mixing truck Mezcladora de concreto sobre camión
Pump truck	Bomba de concreto	
Reinforcement	Refuerzo	
Rebar	Barra de refuerzo, varilla	
Slab	Losa	Slab Losa
Retardant	Retardador	
Additives	Aditivos	
Air-entraining agent	Agente inclusor de aire	
Bonding agent	Agente de vinculación	
Fly ash	Cenizas volantes	Bonding Agent Agente de vinculacion
Pea gravel	Gravita	
Plasticizer	Plasticizador	
Calcium	Calcio	
Quicklime	Cal viva	
Lime putty	Mastique de cal	Pea gravel Gravita

CONCRETE WORK *(cont.)* —
TRABAJO DE CONCRETO *(cont.)*

English	Español	
Clay	Arcilla	
Clod	Gelba	
Curing compound	Compuesta de curado	
Cubic feet	Pies cúbicos	
Yards	Yardas	
Reinforcing steel	Acero para refuerzo	Reinforcing steel Acero para reforzando
Backup bar	Varillas adicionales	
Bar bender	Doblador de varilla	
Rebar caps	Tapas de varilla	
Caps	Tapas	
Keel	Creyón de Madera	
Longitudinal and transverse bars	Varillas longitudinales y transversales	Wire mesh Malla de alambre
Mat	Estera	
Rust	Moho	
Wire mesh	Malla de alambre	
Batter boards	Camilla	
Grade stick	Palo de grado	
Flag stake	Estaca de bandera	Plumb bob Plomo
Pencil	Lápiz	
Plumb bob	Plomo	
String line	Línea de hilo	Pencil Lápiz
Wall line	Línea de la pared	

CONCRETE WORK (cont.) — TRABAJO DE CONCRETO (cont.)

English	Español	
Bottle (oxygen, acetylene)	Botella (oxígeno, acetileno)	
Bottle caps	Tapas de botella	
Cutting torch	Boquilla de corte	
Acetylene	Acetileno	Cutting torch Boquilla de corte
Oxygen	Oxígeno	
Regulator valves	Válvulas de regulación	
Gauges	Medidores, Metros	
Lighter	Encendedor	
Smoke	Humo	
Striker	Percutor	
Tanks	Tanques	
Clearance	Espacio	Gauges Medidores, Metros
Sleeve	Camisa, manga	
Ties	Atadura	
Tie wire	Alambre de atadura	
Wall tie	Ligadura de pared	
Flat ties	Ligadura plana	
Keyway	Llave de cimentación	
Kicker	Puntal	
Valves	Válvulas	
Wire reel	Carrete de alambre	Valves Válvulas
Welding plate	Placa de soldadura	

CONCRETE WORK (cont.) —
TRABAJO DE CONCRETO (cont.)

English	Español	
Wrench	Llave	
Column	Columna	
Pier	Estribo	
Beam	Viga	Wrench Llave
Footing	Zapata de cimentación	
Form	Forma	
Footing form	Forma de zapata de cimentación	
Form oil	Aceite de formas	
Two by four	Dos por cuatro	Footing Zapata de cimentación
Two by twelve	Dos por doce	
Beam forms	Formas de viga	
Stair forms	Formas de escalera	
Planks	Tablones	
Bolt patterns	Muestra de pernos	
Brace	Jabalcon	Form Forma
Bulkhead	Tope de formo	
Turnbuckle	Tensor	
		Two by four Dos por cuatro

CONCRETE WORK (cont.) —
TRABAJO DE CONCRETO (cont.)

English	Español	
Form ply	Madera laminada para formar	
Edge forms	Formas de borde	
Pier and column forms	Formas de estribo y columna	
Slab forms	Formas de losa	
Prestressed concrete	Hormigón prefatigado, hormigón precargado	Panel Panel
Panel	Panel	
Pilaster	Pilastra	
Pins	Pernos	
Shoring	Apuntalamiento	
Curb and gutter	Arroyo encintado	
Driveway	Via de acceso	Pilaster Pilastra
Corner bar	Varilla de esquina	
Curtains	Cortinas	
Dowel	Pasador de varilla	
Embed plate	Placa embutida	Curb and gutter Arroyo encintado
File	Alima	
Waler	Larguero	
Waler loops	Ligaduras de larguero	
Wedge	Calzo	
Z clamp	Grapa en Z	File Alima

CONCRETE WORK (cont.) — TRABAJO DE CONCRETO (cont.)

English	Español	
Darby	Fratás	
Hack saw	Sierra de arco para metal	Hack saw Sierra de arco para metal
Circular saw	Serrucho circular	
Knee pads	Rodilleras	
Rebar shears	Tijeras de varilla	
Boltcutters	Cortapernos	
Boots	Botas	Boots Botas
Face shield	Escudo de cara	
Concrete broom	Escoba de concreto	
Chute	Canal inclinado	
Tool handles (poles)	Mangos de herramientas (tubos)	
Power trowel	Llana mecánica	
Extension cord	Cordón de extensión	Extension cord Cordón de extensión
Stirrup	Brida	
Hickey bar	Dobladora portatil	
Base plate	Placa de base	
Chamfer	Chaflán	
Clamp	Grapa	Expansion joint Junta de expansión
Expansion joint	Junta de expansión	
Control joint	Junta de control	
Cleat	Tirante de formado	

CONCRETE WORK (cont.) —
TRABAJO DE CONCRETO (cont.)

English	Español	
Cone ties	Ligaduras cono de pared	
Corners	Esquinas	Corners
Aggregate fines	Finos de agregado	Esquinas
Sole plate	Placa de base	
Spot footing	Zapata de columna	
Spreader	Seperador	
Dust	Polvo	
Flat work	Pieza plana	
Road base	Base de pavimento	Road base
Water reducer	Reductor de agua	Base de pavimento
Weather	Tiempo (atmosférico)	
Concrete burns	Quemaduras de cemento	
Float	Plana	
Hydration	Hidratación	
Screeds	Botas de goma	Spreader
Screed rod	Formas para pieza plana	Seperador
Slump	Sección	
Vibrator	Recibo del concreto	
Wheelbarrow	Senda de concreto carretilla	
		Wheelbarrow
		Senda de concreto carretilla
Setting time	Tiempo de fraguado	

6-7

English	Español	
D ring	Anillo en D	
Pinch point	Punto de pellizco	
Positioning chain	Cadena de disposición	
Tie off	Atarse seguramente	
Teamwork	Trabajo de cooperación	Cracks
Covering	Recumbrimiento	Rajaduras, grietas
Cracks	Rajaduras, grietas	
Pump	Bomba	
Moist curing	Curación de humedad	
Expansion bolt	Perno de expansión	

D-ring
Anillo en D

Expansion bolt
Perno de expansión

FINISHING AND MASONRY WORK — ACABADAR Y ALBANERIA

English	Español	
Masonry	Albanileria, mamposteria	
Mason	Albañíl	
Brick	Ladrillo	
Firebrick	Ladrillo de fuego	Brick Ladrillo
Face brick	Ladrillo de cara	
Block	Cubos de construcción	
Mortar	Molcajete, mortero	
Reinforced masonry	Mampostería reforzada	
Portland cement	Cemento Portland	
Additives	Aditivos	Block Cubos de construcción
Curing compound	Compuesto para curación	
Silica sand	Arena de silicona	
Grout	Lechada	
Sidewalk	Acera	
Slab	Losa	
Keystone	Clave	Sidewalk Acera
Stairs	Escalones	
Steps	Gradas	
Fireplace	Chimenea	
Control joint	Junta de control	Stairs Escalones

FINISHING AND MASONRY WORK (cont.) — ACABADAR Y ALBANERIA (cont.)

English	Español	
Tools	Herramientas	
Darby	Fratás	
Plumb bob	Plomo	Plumb bob / Plomo
Margin trowel	Cuchara para cemento	
Bullfloat	Plana grande	
Visqueen	Película de polieteno	
Edger	Orillero	Edger / Orillero
Blankets	Mantas	
Chalk line	Rayero de tiza	
Joint	Junta	
Expansion joint	Junta de expansión	
Section	Sección	
Groover	Ranuradora	
Sprayer	Rociador	
Tents	Carpas	
Bucket	Cubo	Bucket / Cubo
Grinder	Moledora	
Grinding wheel	Disco de amoladora	
Hand stone	Piedra de mano	
Neoprene float	Flota de neoprene	Grinder / Moledora

FINISHING AND MASONRY WORK (cont.) — ACABADAR Y ALBANERIA (cont.)

English	Español	
Plastering trowel	Llana de emplastar	
Water brush	Brocha de agua	
Sponge float	Flota de esponga	
Chisel	Cincél	Chisel Cincél
Hammer	Martillo	
Hand jointer	Marcador de juntas de mano	
Dust mask	Mascarilla para polvo	
Fall harness	Arnés de caida	
Laser	Laser	
Grid lines	Líneas de rejilla	Dust mask Mascarilla para polvo
Elevation	Elevación	
Bleedwater	Agua de exudación	
Blistering	Vesiculación	
Sprinkle	Asentamiento	
Acid	Ácido	
Casing nail	Clavo de cabeza perdida	Grid lines Linas de rejilla
Cast stone	Piedra moldeada	
Cavity wall	Muro hueco	
Void space	Espacio vacío	
Coping	Albardilla	Casing nail Clavo de cabeza perdida

ACTIONS — ACCIONES

English	Español
Vibrate	Vibrar
Run (use)	Correr (usar)
Reinforce	Reforza
Drill	Taladar
Grind	Moler
Screw	Atornillar
Shut off	Apagar
Splice	Empalmar
Seal	Sellar
Add water	Agregar agua
Bleed (water)	Exudarse (agua)
Block out	Formar una bloqueda
Adjust	Ajustar
Align	Alinear
Attach	Conectar
Bolt	Meter pernos
Burn in	Pulir
Tie	Atar
Install	Instalar
Repair	Reparar
Scrape	Raspar
Alternate	Alternar
Break	Quebrar
Clip	Esquilar

ACTIONS (cont.) — ACCIONES (cont.)

English	Español
Cut	Cortar
Build	Construir
Check	Comprobar, revisar
Erect	Erigir
Stack	Apilar
Stand	Parar
Coat	Cubrir
Fill	Llenar
Grind	Moler
Mix	Mezclar
Patch	Parchar
Rub	Frotar
Stick	Pegar
Dig	Cavar, excavar
Order	Ordenar
Plan	Planear
Set up	Fraguarse
Lay out	Hacer disposiciones
Mark	Marcar
Shoot (a line)	Tirar (una linea)
Finish	Acabar
Hold	Detener
Bring	Traer

DESCRIPTIONS — DESCRIPCIONES

English	Español
Clean cut	Corte limpio
Center to center	Centro a centro
Tight	Apretado
Level	Nivel
Red hot	Candente
Snap tie	Atadura instantánea
Light	Liviano
Heavy	Pesado
Straight	Derecho
Square	Cuadrado
Thin	Delgado
Wide	Ancho
On grade	De altura
Even	Uniforme
Uneven	Desigual
Wavy	Ondulado
Fresh/green	Fresco/verde
Discolored	Descolorado
Rough finished	Acabado ordinario
Flat	Plano
Rough	Áspero
Smooth	Liso
Honeycombed	Panaleado
High spot	Punto alto
Low spot	Punto bajo

PHRASES — FRASES

English	Español
The fire extinguisher should be kept with the cutting torch.	El extintor se debe guardar con la antorcha de corte.
Alternate ties on this mat.	Alternar las ataduras en esta estera.
Make sure the impalement protection is in place.	Asegúrese que la protección del impalamiento este en lugar.
The short blue flame is the hottest.	La llama azul corta es la más caliente.
Drill a hole in the form.	Taladre un agujero en las formas.
Watch your fingers.	Cuidado con sus dedos.
This wall isn't square.	Esta pared no está cuadrada.
Clean and oil the forms.	Limpie y aceite las formas.
How many yards did you order?	¿Cuántas yardas orede no usted?
Is the preparation all done?	¿Está la prepración totalmete hecha?
Bring your safety glasses, hard hats and tools.	Traiga sus gafas de seguridad, cascos, y herramientas.
Watch the chute.	Este atento con el canal inclinado.
Vibrate over here.	Vibre aquí.
Can you sign this ticket?	¿Puede usted firmar este recibo?

PHRASES *(cont.)* — FRASES *(cont.)*

English	Español
Don't put too much in your wheelbarrow.	No ponga demasiado en su carretilla.
Strip the forms.	Pelar las formas.
Keep your trowel flatter.	Mantenga su llana más aplanada plana.
Feet and inches.	Pies y pulgadas.

CHAPTER 7/CAPITULO 7
Demolition/
Demolición

TOOLS AND EQUIPMENT — HERRAMIENTAS Y EQUIPO		
English	**Español**	
Dump truck	Volquete	
Flat bed truck	Camión de tarima	Dump truck Volquete
Truck bed	Caja de camión	
Truck tailgate	Puerta trasera de camión	
Pick-up truck	Camioneta	
Conveyor	Transportador de mecánico	Pick-up truck Camineta
Dragline	Excavadora de arrastre	
Dredge	Draga	
Cart	Carreta	
Chain	Cadena	Chain Cadena
Jack	Gato	
Hose	Manguera	

TOOLS AND EQUIPMENT (cont.) — HERRAMIENTAS Y EQUIPO (cont.)

English	Español	
Jackhammer	Martillo neumático	
Pry bar	Barra de palanca	Pry bar Barra de palanca
Shovel	Pala	
Sledgehammer	Almadena	
Wheel barrow	Carretilla	Shovel Pala
Torch	Antorcha	
Chute	Dúcto	
Rubbish chute	Dúcto de basura	Wheel barrow Carretilla

OTHER DEMOLITION — OTRA DEMOLICIÓN

English	Español	
Debris	Ruina	
Dump	Basurero	
Dust	Polvo	
Garbage	Basura	
Rubble	Escombro	
Glass	Vidrio	
Land fill	Deposito de basura	
Metal pile	Pila de metal	
Abatement	Anulación (remoción)	
Contamination	Contaminación	

Garbage
Basura

Rubble
Escombro

ACTIONS — ACCIONES

English	Español
Blast	Ráfaga
Dynamite	Dinamitar
Demolish	Demoler
Destroy	Destruir
Break	Romper
Hit	Pegar
Chip	Picar
Pound	Apilar
Ram	Impelar
Knock over	Tumbar
Smash	Allanar
Tear apart	Desarmar
Push	Empujar
Pull	Jalar
Collect	Recoger
Haul	Transportar
Dump	Descargar
Cover	Cubrir
Fold	Doblar
Keep	Guardar

ACTIONS (cont.) — ACCIONES (cont.)	
English	**Español**
Pile	Apilar
Recycle	Reciclar
Remove	Quitar
Salvage	Salvar
Scoop	Recoger
Scrap	Desechar
Throw away	Descartar
Collapse (structure)	Colapsó

DESCRIPTIONS — DESCRIPCIONES

English	Español
Dusty	Polvoriento
Explosive	Explosivo
Protruding	Saliente
Sharp	Filoso
Strong	Fuerte
Stuck	Travado
Weak	Débil
Junk	Junce, chatarra
Unsafe building	Edificio inseguro, edificación insegura

PHRASES — FRASES

English	Español
Salvage any metal that you find.	Salve caulquier metal que encuentre.
Watch out for protruding nails.	Fijese por clavos salidos.
Stand back while I break this glass.	Manténgase alejado mientras rompo esta cristal.
Don't go in that area, it's secured for asbestos removal.	No entre en esa área, esta asegurada por el retiro de asbestos.
Wear these dust masks.	Usen estas mascarillas para el polvo.

NOTES — NOTAS

CHAPTER 8/CAPITULO 8
Welding
Soldadura

WELDING — SOLDADURA		
English	*Español*	
Welder	Soldador	
Welding rod	Electrodo	
AC current	Corriente CA	
DC current	Corriente CC	
Arc welding	Soldadura al arco	Welder Soldador
Amperes	Amperios	
Amperage	El amperaje	
Volts	Voltios	
Arc	Arco	Volts Voltios
Ground lead	Cable conductor de tierra	
Welding lead	Cable conductor de soldadura	
Filler rod	Barra rellenadora	
Welding apron	Delantal de soldadura	
Welding gloves	Guantes de soldadura	Welding apron Delantal de soldadura
Welder's vest	Chaleco de soldadura	
Wire feed welder	Soldadura de alimentación de alambre	
Puddle weld	Soldadura de charco	Welding gloves Guantes de soldura

WELDING (cont.) — SOLDADURA (cont.)

English	Español	
Flush weld	Soldadura a lísa	
Lap weld	Soldadura a solape	
Butt weld	Soldadura a tope	
Flat weld	Soldadura de plano	
Overhead weld	Soldadura de encima	
Vertical weld	Soldadura vertical	
Fillet weld	Soldadura orthogonal	Hand signal / Señal de mano
V joint	Junta en V	
Flux	Fundente	
Hand signal	Señal de mano	
Sparks	Chispas	
Heat	Calor	
Flash burns	Quemadura de relámpago	
Burns	Quemaduras	
Welding blanket	Manta de soldadura	Fire extinguisher / Extintor
Fall protection	Protección	
Fire extinguisher	Extintor	
Scissor lift	Plataforma hidráulica	

Hand signal
Señal de mano

Fire extinguisher
Extintor

Scissor lift
Plataforma hidráulica

WELDING (cont.) — SOLDADURA (cont.)

English	Español	
Gauges	Indicadores	Gauges Indicadores
Acetylene	Acetileno	
Oxygen	Oxígeno	
Magnesium	Magnesio	
Bead	Cordón	
Buildup	Acumulación	
Butt joint	Junta de tope	
Hand grinder	Afilador de mano	C clamp Grapa en C
C clamp	Grapa en C	
Vise	Tornillo de banco	
Wire brush	Cepillo de alambre	
Slag hammer	Martillo de escoria	
Joint	Junta	
Overlap	Traslapo, encimar	
Rust	Moho	Vise Tonníllo de banco
Slag	Escoria	
Steel	Acero	
X-ray	Radiografía	
Penetration	Penetración	Steel Acero

ACTIONS — ACCIONES

English	Español
Weld	Soldar
Ground	Conectar a tierra
Cut	Cortar
Grind (metal)	Afilar (moler)
Brush	Cepillar
Check	Comprobar
Cool	Enfriar
Insert	Insertar
Inspect	Examinar, inspeccionar
Preheat	Precalentar
Secure	Asegurar
Burn	Quemar

DESCRIPTIONS — DESCRIPCIONES

English	Español
Even	Uniforme
Uneven	Desigual
Tight	Apretado
Loose	Flojo
Continuous	Continuo
Securely	Seguramente
Cracked	Agrietado
Defective	Defectuoso
Distorted	Torcido
Ductility	Ductilidad
Fabricated	Fabricado
Galvanized	Galvanizado
Greasy	Grasiento, grasoso
Jammed	Atorado
Reverse threaded	Reverso roscado
Rough	Áspero
Shiny	Brilliante
Smooth	Liso
Stuck	Pegado
Voltage	Voltaje
Puddle	Charco
Red hot	Fosforescente

PHRASES — FRASES

English	Español
Strike an arc.	Prender un arco.
Don't look at the arc.	No mire al arco.
Is the welding lead connected well?	¿Está bien conectado el conductor de soldadura?
That joint is still hot.	Ese empalme toda vía está caliente.
Keep your bead even.	Mantenga su cordón uniforme.

CHAPTER 9/CAPITULO 9
Finishing
Acabado

ROOFING — TECHADO		
English	*Español*	
Roof	Techo	
Roofer	Techero	
Dead load	Carga muerta	
Deck, decking	Cubierta	
Metal deck	Plataforma metálica	
Kettle	Caldera	
Felt	Fieltro	
Tar	Alquitran, brea, chapopote	
Tar paper	Papel de brea	
Asphalt	Asfalto	
Roofing felt	Tela asfaltica, felpa	
Flashing	Cubrejuntas, tapajuntas	
Curb	Guarnición	
Hatch	Compuerta	
Skylight	Tragaluz, claraboya	
Sheet metal	Lámina metálica, chapa metálica	
Sheet copper	Lámina de cobre	
Roof deck	Cubierta de techo	

Roof
Techo

Curb
Guarnicion

Tar paper
Papel de brea

Skylight
Tragaluz, claraboya

English	Español	
Roof drain	Desagüe de techo	
Flat roof	Techo plano	
Roof covering	Revestimiento de techo, cubierta de techo	
Wood shake (shingle)	Teja de Madera, ripia	
Shingle	Teja	Roof drain / Desague de techo
Asphalt shingle	Teja de asfalto	
Wood shingle	Ripia	
Slate shingle	Teja de pizarro	
Roof tile	Teja	
Roof sheathing	Entarimado de teja	Shingle / Teja
Ridge board	Tabla de cumbrera	
Rafter	Cabrio, cabio	
Ridge tile	Tejas para cumbrera	
Gable	Hastial	
Gable roof	Techo a dos aguas	Gable / Hastial
Hip	Lima	
Hip roof	Techo a cuatro aguas	
Mansard roof	Mansardara	Mansard roof / Mansardara

ROOFING *(cont.)* — TECHADO *(cont.)*

English	Español	
Overhang	Voladizo, vuelo, alero	
Ridge	Cresta	
Roofing square	Cuadro de cubierta de techo	
Dormer	Buharda	
Eave	Alero	
Gutter	Gotera	
Uplift (wind)	Remonte	
Waterproofing	Impermeable	
Leak	Fugs, gotera	
Ceiling	Cielo	
Utility knife	Navaja de utilidad	
Caulk	Calafatear	
Sealant	Sellador	

Ridge
Cresta

Gutter
Gotera

Utility knife
Navaja de utilidad

FLOORING — ENTARIMADO

English	Español	
Flooring	Revestimiento para pisos	
Wood flooring	Piso de Madera	
Underlayment	Suelo subyacente	
Interlayment	Capa intermedia	
Floor tiles	Baldosas	
Carpet	Alfombra	
Carpet layers	Alfombreros	
Ceramic tile	Azulejos ceramicos	
Kraft paper	Papel Kraft	
Tile setter	Azulejero	
Adhesive	Pegamento	
Threshold	Umbral	
Drain	Desagüe	
Utility knife	Navaja de utilidad	

Carpet
Alfombra

Tile setter
Azulejero

Threshold
Umbral

EXTERIOR — EXTERIOR

English	Español	
Siding	Forrado, revestimiento	
Vinyl siding	Revestimiento de vinilo	
Aluminum siding	Revestimiento de aluminio	
Lap siding	Revestimiento de tablas con traslape	Vinyl siding Revestimiento de vinilo
Stucco	Estuco	
Facing brick	Ladrillos para frentes	
Seasoned wood	Madera de estacionada	
Exit door	Puerta de salida	
Insulation	Aíslamiento	
Facade	Alzado	
Glazing	Vidriado	Exit door Puerta de salida
Safety glazing	Vidriado de seguridad	
Awnings	Toldos	
Soffit	Sofito	
Fascia	Omposta	
Adhesive	Pegamento	
		Glazing Vidriado

English	Español	
Caulking	Calafeto	
Caulking gun	Pistola de calafeto	
Sealant	Sellador	
Gate	Puerta de cerco	
Tree	Árbol	
Sod	Césped	
Sprinkler control box	Caja de controles de regadera	
Sprinklers	Regaderas	

Caulking gun
Pistola de calafeto

Tree
Árbol

INTERIOR — INTERIOR

English	Español	
Drywall	Pirca	
Sheetrockers	Yeseros	
Gypsum	Yeso	
Gypsum board	Panel de yeso	
Taping compound	Pasta de muro	
Plaster	Azotado, jaharro	
Plastering	Revoque, enclucido	
Lath	Liston	
Furring, furred	Enrasado	
Trim	Molduras	
Joint compound	Pasta de muro	
Folding partition	Tabique plegable	
Moving partition	Tabique movable	
Portable partition	Tabique portátil	
Handrails	Barandales	
Ornamental metal	Metal ornamental	
Ornamental railing	Pasamanos ornamental	
Veneer	Revestimiento	
Baseboard	Rodapie	
Molding	Moldura	
Casing	Chambrana	
Reglet	Regleta	
Cornice	Cornisa	

Drywall
Pirca

Joint compound
Pasta de muro

Handrails
Barandales

Casing
Chambrana

INTERIOR (cont.) — INTERIOR (cont.)

English	Español	
Sill plate	Solera inferior	
Paneling	Empalenado	
Tongue and groove	Machihembrado	
Wainscoting	Friso, alfarje	
Rail	Cremallera barandal	
Railing	Barandal, barra, carril	Railing Barandal, barra, carril
Wood putty	Masilla para Madera	
Putty knife	Espátula de masilla	
Sand paper	Lija	
Spackle	Junta de cemento	Putty knife Espátula de masilla
Cabinets	Gabinetes	
Cabinetmaker	Ebanista	
Shelf	Repisa	
Latch	Aldaba	
Face grain	Veta superficial	
Grille	Rejilla	
Toe board	Tabla de pie	Cabinets Gabinetes
Runners	Largueros	
Swinging door	Puerta pivotante	
Door	Puerta	
Door closer	Freno de puerta	
Door stop	Tope	Door Puerta

English	Español	
Hinges	Bisagras	 Hinges Bisagras
Key	Llave	
Lock	Cerradura	
Door stop	Tope	
Window	Ventana	 Window Ventana
Painter	Para	
Paint brush	Brocha para pintura	
Paint roller	Rodillo para pintura	
Primed	Imprimado	
Primer	Imprimador	
Putty coat	Enlucido	 Paint brush Brocha de pintura
Pour coat	Capa de colada	
Utility knife	Navaja de utilidad	
Sponge	Esponja	
Masking tape	Cinta de enmascarar	 Paint roller Rodillo de pintura
Caulking	Calafeto	
Caulking gun	Pistola de calafeto	
Cloth	Tela	
Adhesive	Pegamento	
Caulking	Calafeto	 Caulking gun Pistola de calafeto
Ceiling tiles	Tejas de cielo	
Planter	Plantador	
Plants	Plantas	

INTERIOR *(cont.)* — INTERIOR *(cont.)*

English	Español	
Spiral stairs	Escaleras de caracol	
Stairwell	Recinto de escaleras	
Stud finder	Buscador be montantes	
Templates	Plantillas	
Texture	Textura	
Power doors	Puertas mecánicas	
Revolving door	Puerta giratoria	

Spiral stairs
Escaleras de caracol

ACTIONS — ACCIONES	
English	*Español*
Install	Instalar
Run (use)	Correr (usar)
Extend	Extender
Modify	Modificar
Mark	Marcar
Measure	Medir
Fasten	Atar
Anchor, affix	Anclar
Manufacture	Fabricar
Verify	Verificar
Screw	Atornillar
Shut off	Apagar
Test	Probar
Record	Registrar
Submit	Someter
Lift	Levantar
Lower	Bajar
Roll	Rodar
Put on	Poner
Locate	Localizar
Spray	Rociar

ACTIONS *(cont.)* — ACCIONES *(cont.)*

English	Español
Wash	Lavar
Check	Comprobar
Warn	Advertir
Ventilate	Ventilar
Splice	Empalmar
Finish	Acabar
Hold	Detener
Bring	Traer
Leak	Gotear
Lock	Cerrar (bajo llave)
Paint	Pintar
Patch	Parchar
Plant	Plantar
Position	Posicionar
Press	Prensar
Sand	Lijar
Seal	Sellar
Seed	Sembrar
Swing	Oscilar
Adjust	Ajustar
Caulk	Calafatear
Complete	Cumplir
Hang	Colgar

ACTIONS (cont.) — ACCIONES (cont.)

English	Español
Install	Instalar
Lay (carpet)	Poner (alfombra)
Open	Abrir
Paint	Pintar
Reinforce	Reforzar
Replace	Reemplazar

DESCRIPTIONS — DESCRIPCIONES

English	Español
Ready	Listo
Safely	Con seguridad
Slowly	Lento
Approved	Aprobado
Rejected	Rechazado
Interior	Interior
Tight	Apretado
Level	Nivel
Straight	Derecho
Square	Cuadrado
Exterior	Exterior
Underground	Subterráneo
Located	Localizado
Parallel	Paralelo
Perpendicular	Perpendicular
Heavy	Pesado
Loose	Flojo
Center to center	Centro a centro

DESCRIPTIONS *(cont.)* — DESCRIPCIONES *(cont.)*

English	Español
Missing	Falta
Done	Hecho
Flush	A nivel
Exterior	Exterior
Finished	Acabado
Interior	Interior
Nicked	Muescado

NOTES — NOTAS

CHAPTER 10/CAPITULO 10
Plumbing
Plomería

PLUMBING — PLOMERÍA		
English	**Español**	
Plumbing	Instalación de hidráulicas y sanitarias, plomería	Plumbing plan Plano de plomería
Plumbing appliance	Mueble sanitario	
Plumber	Plomero	
Plumbing plan	Plano de plomería	
Pipe, piping	Cañería, caño, tubo	
Manhole	Pozo de confluencia, boca de inspección, boca de accesso, pozo de entrada	Manhole Pozo de confluencia boca de inspeccion boca de acceso pozo de entrada
Main vent	Respiradero matriz	
Main	Principal, matriz	
Leader (pipe)	Tubo de bajada	
Lateral (pipe)	Ramal lateral	
Vertical pipe	Tubo vertical	
Branch	Ramal	Elbow Codo
Elbow	Codo	

10-1

PLUMBING *(cont.)* — PLOMERÍA *(cont.)*

English	Español	
Thread	Hilo	
Tee	T, injerto	
Radius	Radio	
Diameter	Diámetro	
Valve	Válvula	
Cut-off valve	Válvula de cierre	
Check valve	Válvula de contraflujo	
Bleeder valve	Válvula de purga	
Hub valve	Válvula de cubo	
Key valve	Válvula de llave	
Relief valve	Válvula de alivio, llave de alivio	
Cleanout	Registro	
Ball cock	Válvula de flotador	
Ball valve	Llave de flujo	
Grease interceptor	Interceptor de grasas	
Grease trap	Collector de grasas	

Tee
T, injerto

Radius

Radius
Radio

Valve
Válvula

Cleanout
Registro

English	Español	
Trap	Sifón	
Trap seal	Sello de trampa hidráulica	
Urinal	Orinal	**Trap** Sifón
Toilet	Inodoro, sanitario	
Faucet	Llave	
Spigot	Llave, grifo, canilla	
Pump	Bomba	
Sump	Sumidero	**Toilet** Inodoro, sanitario
Sump pump	Bomba de sumidero	
Sump vent	Respiradero de sumidero	
Shower stall	Ducha, regadera	
Showerhead	Regadera	**Pump** Bomba
Hose bibb	Grifo de manguera	
Shutoff valve	Válvula de cierre	
Sill cock	Grifo de manguera	
Wax seal	Empaque de cera	
To vent	Ventilar, evacuar	**Showerhead** Regadera

English	Español	
Vent shaft	Recinto de ventilación	
Vent system	Sistema de ventilación	
Vent stack	Respiradero vertical	
Venting system	Sistema de evacuación	Venting system Sistema de evacuación
Standpipe	Columna hidrante	
Standpipe system	Sistema de columna hidrante	
Stack	Tuberia vertical bajante	
Stack vent	Respiradero de bajante	
Soil pipe	Tubo bajante de aguas negras	Sprinkler head Rociador
Soil stack	Bajante sanitario	
Sprinkler	Rociador	
Sprinkler head	Rociador	
Sprinkler system	Sistema de rociadores	
Torch	Antorcha	
Brass	Bronce	
Braze	Soldar en fuerte	
Brazing alloy	Aleación para soldar	Torch Antorcha

PLUMBING (cont.) — PLOMERÍA (cont.)

English	Español	
Brazing flux	Fundente para soldar	
Propane	Propano	Propane / Propano
Cistern	Aljibe	
Chase	Canaletas	
Bracket	Brazo	
Backing	Soporte	
Backflow	Contraflujo	
Bathroom	Cuarto de baño	
Bathroom sink	Lavabo	
Bathtub	Bañera	
Septic tank	Fosa séptica	Hydrant / Boca de riego
Well (water)	Aljibe, pozo de agua	
Washer and dryer	Lavadora y secadora	
Hydrant	Boca de riego	
Templates	Plantillas	
To Bury	Enterrar	
Plunger	Destapacaños, sopapa	
Channel-lock pliers	Alicates de extensión	Plunger / Destapacaños, sopapa

PLUMBING (cont.) — PLOMERÍA (cont.)

English	Español	
Vise-grip pliers	Alicates de presión, pinzas perras	
Clamp	Grapa, abrazadera	
Level	Nivel	
Plumb bob	Plomo	
Basin wrench	Llave pico de gansa	
Chain wrench	Llave de cadena	
Pipe wrench	Llave de tubo	
Valve-seat wrench	Llave de asiento de válvula	
Snake	Serpiente	
Vise	Morsa	
Vice bench	Torno de banco	

Plumb bob
Plomo

Pipe wrench
Llave de tubo

ACTIONS — ACCIONES

English	Español
Install	Instalar
Run (use)	Correr (usar)
Extend	Extender
Modify	Modificar
Mark	Marcar
Measure	Medir
Fasten	Atar
Bury	Enterrar
Anchor, affix	Anclar
Manufacture	Fabricar
Verify	Verificar
Screw	Atornillar
Shut off	Apagar
Test	Probar
Record	Registrar
Submit	Someter
Lift	Levantar
Lower	Bajar
Roll	Rodar
Put on	Poner
Locate	Localizar
Spray	Rociar
Wash	Lavar

ACTIONS (cont.) — ACCIONES (cont.)

English	Español
Check	Comprobar
Warn	Advertir
Ventilate	Ventilar
Splice	Empalmar
Seal	Sellar
Finish	Acabar
Hold	Detener
Bring	Traer
Lock	Cerrar

DESCRIPTIONS — DESCRIPCIONES

English	Español
Ready	Listo
Safely	Con seguridad
Slowly	Lento
Approved	Aprobado
Rejected	Rechazarado
Interior	Interior
Tight	Apretado
Level	Nivel
Straight	Derecho
Square	Cuadrado
Exterior	Exterior
Underground	Subterráneo
Located	Localizado
Parallel	Paralelo
Perpendicular	Perpendicular
Heavy	Pesado
Loose	Flojo
Center to center	Centro a centro
Missing	Le falta

NOTES — NOTAS

CHAPTER 11/CAPITULO 11
HVAC and Mechanical
HVAC y Mecanico

HVAC AND MECHANICAL — HVAC Y MECANICO		
English	*Español*	
HVAC	Calefacción, ventilacion y aire acondicionado	
Mechanical plan	Plano mecánico	
Heating	Calefacción	
Heater	Calefactor, estufa	Mechanical plan
Natural gas	Gas natural	Plano mecánico
Gas main	Conducto principal de gas	
Self-ignition	Auto-ignición	
Boiler	Caldera	
Boiler room	Sitio de la caldera	Gas main
Pipes	Pipas	Conducto principal de gas
Elbow	Codo	
Radius	Radio	
Diameter	Diámetro	
Steam	Vapor	Elbow
Hot water	Agua caliente	Codo
Furnace	Horno	
Flue	Conductos de humo	Radius
Lining	Recubrimiento	Radius / Radio

HVAC AND MECHANICAL *(cont.)* — HVAC Y MECANICO *(cont.)*

English	Español	
Chimney	Chimenea	
Chimney liner	Revestimiento de chimenea	
Cleanout (chimney)	Abertura de limpieza	Chimney Chimenea
Coal	Hulla	
Air conditioning	Aire acondicionado	
Coolant	Líquido refrigerante	
Cooling tower	Torre de refrigerado	
Backflow	Contraflujo	
Insulation	Aislamiento/la insulación	Air conditioning Aire acondicionado
Duct	Conducto	
Duct worker	Trabajador del conducto	
Fan	Ventilador	
Exhaust	Escape, extracción	
Damper	Regulador	
Dead end	Terminal, sin salida	Fan Ventilador
Hangers	Ganchos	
Hood (kitchen)	Campana (cocina)	
Louver	Celosia	
Plenum	Pleno, camara de distribución de aire	
Smoke	Humo	
Smoke barrier	Barrera antihumo	Smoke Humo

HVAC AND MECHANICAL (cont.) — HVAC Y MECANICO (cont.)

English	Español	
Smoke-tight	Impermeables al humo	
Suspended ceiling	Falso plafón, cieloraso suspendido	
Acoustical panels	Paneles acústicos	
Vacuum	Vacío, aspiradora	Thermostat / Termostato
Thermostat	Termóstato	
Setback	Retiro	
Access	Acceso	
Access cover	Tapa de acceso	
Automatic closing device	Dispositivo de cierre automático	Access cover / Tapa de acceso
Chase	Canaleta	
Sleeve	Camisa, manga	
Meter (measuring device)	Metro, medidor	
Gauge (instrument)	Monometro	Gauge (instrument) / Monometro
Pressure	Presión	
Regulator	Regulador	
Ice	Hielo	
Maintenance	Mantenimiento	
Manufacturer	Fabricante	
Templates	Plantillas	Ice / Hielo

HVAC AND MECHANICAL *(cont.)* — HVAC Y MECANICO *(cont.)*

English	*Español*	
Factory	Fábrica	Factory Fábrica
Mechanic	Mecánico	
Machinist	Maquinista	
Rough-in	Instalación en obra negra, instalación de obra gruesa	
Battery	Batería	Battery Pila
Alarm	Alarma	
Fire alarms	Alarma de incendio	
Fire sprinklers	Regaderas de fuego	
Fire proofing	Ignifugación	
Warranty	Garantía	Alarm Alarma
Subcontractor	Subcontratista	

ACTIONS — ACCIONES

English	Español
Install	Instalar
Run (use)	Correr (usar)
Extend	Extender
Modify	Modificar
Mark	Marcar
Measure	Medir
Fasten	Atar
Bury	Enterrar
Anchor, affix	Anclar
Manufacture	Fabricar
Verify	Verificar
Screw	Atornillar
Shut off	Apagar
Test	Probar
Record	Registrar
Submit	Someter
Lift	Levantar
Lower	Bajar
Roll	Rodar
Put on	Poner
Locate	Localizar
Spray	Rociar
Wash	Lavar

ACTIONS (cont.) — ACCIONES (cont.)	
English	*Español*
Check	Comprobar
Warn	Advertir
Ventilate	Ventilar
Splice	Empalmar
Seal	Sellar
Finish	Acabar
Hold	Detener
Bring	Traer
Lock	Cerrar

DESCRIPTIONS — DESCRIPCIONES

English	Español
Ready	Listo
Safely	Con seguridad
Slowly	Lento
Approved	Aprobado
Rejected	Rechazado
Interior	Interior
Tight	Apretado
Level	Nivel
Straight	Derecho
Square	Cuadrado
Exterior	Exterior
Underground	Subterráneo
Located	Localizado
Parallel	Paralelo
Perpendicular	Perpendicular
Heavy	Pesado
Loose	Flojo
Center to center	Centro a centro
Missing	Le falta

NOTES — NOTAS

CHAPTER 12/CAPITULO 12
Electrical
Electricista

ELECTRICAL — ELECTRICISTA		
English	**Español**	
Electrician	Electricista	
Electricity	Electricidad	
Voltage	Voltaje	
Volts	Voltios	
AC current	Corriente CA	
DC current	Corriente CC	Electrician Electricista
Amperes	Amperias	
Amperage	El amperaje	
Resistance	Resistencia	
Impedance	Impedencia	
Power	Potencia	Volts Voltios
Frequency	Frecuencia	
Series	Serie	
Parallel	Paralelo	
Transformer	Transformador	
Circuit	Circuito	$R_T = 10\Omega$ Series Serie

ELECTRICAL (cont.) — ELECTRICISTA (cont.)

English	Español	
Phase	Fase	Parallel Paralelo
A phase, B phase, C phase	Fase 1, fase 2, fase 3	
Neutral	Neutro, neutral	
Neutral conductor	Cable neutro	
Electrical plan	Plano eléctrico	
Copper	Cobre	
Hard drawn copper	Cobre estirado en frío	Transformer Transformador
Aluminum	Aluminio	
Feeder	Alimentador	
Conductor	Conductor	
Insulation	Aislamiento/la insulación	
Wire	Alambre	Electrical plan Plano eléctrico
Connection	Conexión, unión	
Splice	Empalme, translape, junta	
		Wire Alambre

ELECTRICAL (cont.) — ELECTRICISTA (cont.)

English	Español	
Buried cable	Cable enterrado	
Bury	Enterrar	
Power lines	Líneas de energía	
Underground lines	Líneas subterráneas	
Manhole	Pozo de confluencia, boca de inspección, boca de accesso, pozo de entrada	Power lines Líneas de energía
Feeder cable	Cable de alimentación	
Main power cable	Cable principal	
Service entrance neutral	Cable principal neutro	
Wire connectors	Conectores de alambre, cable alambre conector	Wire connectors Conectores de alambre, cable alambre conector
Conduit	Conducto, canal	
Raceway	Conducto eléctrico	
Enclosure	Cerramiento	
Pipe	Pipa	
Elbow	Codo	
Radius	Radio	Elbow Codo
		Radius Radius Radio

ELECTRICAL (cont.) — ELECTRICISTA (cont.)

English	Español	
Diameter	Diámetro	
Cable tray	Bandeja de portacables	
Connector	Conector	Diameter / Diametro
Screw connector	Conector con tornillo	
Coupling	Acoplamiento	
Fitting	Accesorio	
Flexible conduit	Conducto portacables flexible	
Outlet box	Caja de enchufe, Caja de tomacorriente	Screw connector / Conector con tornillo
Junction box	Caja de conexiones de empalme	
Cabinets	Gabinetes	
Opening	Abertura	
Knockout	Agujero ciego	
Hangers	Ganchos	Outlet box / Caja de enchufe, Caja de tomacorriente
Riser (pipe)	Tubo vertical	
Chase	Canaletas	
Plenum	Pleno, cámara de distribución de aire	
Dead end	Terminal, sin salida	Junction box / Caja de conexiones de empalme

ELECTRICAL (cont.) — ELECTRICISTA (cont.)

English	Español	
Offset	Desplazamiento, desvío	
Rough-in	Instalación en obra negra, instalación de obra gruesa	Offset Desplazamiento, desvío
Main	Principal, matriz	
Main breaker	Interruptor automático principal	
Circuit breaker	Apagador, interruptor de circuito	
Circuit breaker panel	Cuadro de cortacircuito	
Single pole breaker	Interruptor automático unipolar	Circuit breaker Apagador, interruptor de circuito
Double pole breaker	Interruptor automático bipolar	
Vacuum breaker	Interruptor de vacío	
Cartridge fuse	Fusible de cartucho	
Fuse	Fusible	
Fuse box	Caja de fusibles	
Plug fuse	Fusible de rosca	
Disconnect switch	Desconectar	Circuit breaker panel Cuadro de cortacircuito
Bonding conductor	Cable de enlace	
Bonding jumper	Borne de enlace	
Ground connection	Conexión de tierra	
		Fuse Fusible

12-5

English	Español	
Ground bar	Bandeja a tierra	
Ground rod	Barra a tierra, vara a tierra	
Neutral bar	Bandeja neutral	Ground rod
Ground wire	Cable a tierra	Ground rod Barra a tierra, vara a tierra
Hot bus bar	Bandeja de carga	
Electrical fixture	Artefactos eléctricos	
Electrical outlet	Enchufe, tomacorriente	
Range outlet	Tomacorriente para estufa, enchufe para estufa	Fluorescent Fluorescente
Fixture	Artefacto	
Fluorescent	Fluorescente	
Ballast	Lastre	
Floodlight	Iluminación	
Self-luminous	Autoluminoso	Light bulb Foco
Lamp	Lámpara	
Light bulb	Foco	
Lights	Luces	
Ground fault circuit interruptor (GFCI)	Interruptor fusible de seguridad a tierra	Reset button Test button
Loop	Lazadas	Ground fault circuit interruptor (GFCI) Interruptor fusible de seguridad a tierra
Plastic insulator	Aislante plástico	

ELECTRICAL (cont.) — ELECTRICISTA (cont.)

English	Español	
Receptacle, plug	Enchufe, clavija	
Switch	Interruptor, apagador	
Switch plate	Placa de interruptor	
Automatic closing device	Dispositivo de cierre automático	Receptacle, plug Enchufe, clavija
Self-closing	Autocierre	
Interlocking	Enclavamiento	
Motor	Motor	
Fan	Ventilador	
Exhaust fan	Ventilador de extracción	
Exhaust	Escape, extracción	Switch Interruptor, apagador
Access	Acceso	
Access cover	Tapa de acceso	
Fire alarm	Alarma de incendio	
Fire sprinklers	Regaderas de fuego	
Fire proofing	Ignifugación	
Smoke detector	Sensor de humo	Motor Motor
Manual pull station	Alarma de incendio manual	
Fumes	Gases	
Thermostat	Termóstato	
		Smoke detector Sensor de humo

ELECTRICAL (cont.) — ELECTRICISTA (cont.)

English	Español	
Battery	Pila	
Alarm	Alarma	
Telephone	Teléfono	
Communication system	Sistema de comunicación	Battery Pila
Security system	Sistema de seguridad	
Television system	Sistema de televisión	
Rack	Caja de controles portátil	
Door bell	Timbre	
Meter (measuring device)	Metro, medidor	Telephone Teléfono
Code	Código	
Generator	Generador	
Open air	Aire libre	
Power strip	Zapatilla eléctrica	
Power supply	Fuente de alimentación	Meter (measuring device) Metro, medidor
Premises	Local, sitio	
Show window	Vitrina	
Showcase	Armario de exhibición	
Temporary	Provisional	
Temporary power	Energía temporaria	
Sleeve	Camisa, manga	
Sparks	Chispas	Show window Vitrina

English	Español	
Extension cord	Cable de extensión	Extension cord / Cable de extensión
Responsibility	Responsabilidad	
Evacuate	Evacuar	
Lockout/tagout	Cierre	
Unbalanced loads	Cargas no balanceadas	
No smoking area	Zona de no fumar	
Warning	Aviso	
Live wires	Alambres vivos	
Suspended ceiling	Falso plafón, cieloraso suspendido	Lockout/tagout / Cierre
Acoustical panels	Paneles acústicos	
Ducts	Conductos	
Elevator	Elevador	
Factory	Fábrica	
Manufacturer	Fabricante	
Templates	Plantillas	Factory / Fábrica
Tool lockup	Cuarto de herramientas bajo llave	
Maintenance	Mantenimiento	
Mechanic	Mecánico	
Warranty	Garantía	
Subcontractor	Subcontratista	
Foreman	Capataz	Foreman / Capataz

ACTIONS — ACCIONES

English	Español
Install	Instalar
Run (use)	Correr (usar)
Extend	Extender
Modify	Modificar
Mark	Marcar
Measure	Medir
Fasten	Atar
Anchor, affix	Anclar
Manufacture	Fabricar
Verify	Verificar
Screw	Atornillar
Shut off	Apagar
Test	Probar
Record	Registrar
Submit	Someter
Lift	Levantar
Lower	Bajar
Roll	Rodar
Put on	Poner
Locate	Localizar
Spray	Rociar
Wash	Lavar
Check	Comprobar

ACTIONS (cont.) — ACCIONES (cont.)

English	Español
Warn	Advertir
Ventilate	Ventilar
Splice	Empalmar
Seal	Sellar
Finish	Acabar
Hold	Detener
Bring	Traer
Lock	Cerrar
Charge (batteries)	Cargar (las pilas)

DESCRIPTIONS — DESCRIPCIONES

English	Español
Ready	Listo
Safely	Con seguridad
Slowly	Lento
Approved	Aprobado
Rejected	Rechazado
Interior	Interior
Tight	Apretado
Level	Nivel
Straight	Derecho
Square	Cuadrado
Exterior	Exterior
Underground	Subterráneo
Located	Localizado
Parallel	Paralelo
Perpendicular	Perpendicular
Heavy	Pesado
Loose	Flojo
Center to center	Centro a centro
Missing	Falta

CHAPTER 13/CAPITULO 13
Safety
Seguridad

SAFETY — SEGURIDAD		
English	**Español**	
911 (Nine-one-one)	911 (Nueve-uno-uno)	
Accident	Accidente	
Emergency	Emergencia	
Ambulance	Ambulancia	
Hospital	Hospital	
Clinic	Clínica	911 (Nine-one-one) 911 (Nueve-uno-uno)
Injury	Lesión	
Sick	Enfermo	
Hurt	Herido	
Pain	Dolor	
Conscious	Conciente	Ambulance Ambulancia
Cut	Cortada	
Collapse (person)	Colapaso	
Drunk	Borracho	Cut Cortada
Healthy	Sano	
Insurance	Aseguranza	
Rules	Reglas	
Inspection	Inspección	
Preparation	Preparación	Hospital Hospital

SAFETY (cont.) — SEGURIDAD (cont.)

English	Español	
Safety policy	Poliza de seguridad	
Documentation	Documentación	
Disinfectant	Desinfectante	
Sanitation	Saneamento sanidad salubridad	Warning signs Señales de peligro
Hazard	Peligro	
Warning signs	Señales de peligro	
Lethal	Mortal	
Lockout/tagout	Cierre	
No smoking area	Zona de no fumar	
Responsibility	Responsabilidad	
Training	Entrenamiento	
Lockout/tagout	Cierre	Lockout/tagout Cierre
Flammable liquid	Líquido inflamable	

MEDICAL — MEDICINA

English	Español	
Antibiotic	Antibiótico	
Poison	Veneno	
Sterile	Estéril	
Bandage	Vendaje	
Splint	Tabilla	
Tetanus shot	Inyección contra el tetanus	Bandage Vendaje
Dehydrated	Deshidratado	
Heat stroke	Insolación	
Medicine	Medicación	
Medicine cabinet	Botiquín	
		Medicine Medicación

HAZARDS — PELIGROS

English	Español	
Live wires	Alambres vivos	
Paint thinners	Diluyentes de pintura	
Gasoline	Gasolina	
Sliver	Astilla	
Asbestos	Asbesto	
Fire	Fuego	

Gasoline
Gasolina

Fire
Fuego

SAFETY EQUIPMENT — EQUIPO DE SEGURIDAD

English	Español	
Baricade	Barricada	
Head protection	Protección para la cabeza	Goggles Anteojos
Gloves	Guantes	
Eye protection	Protección de ojos	
Ear plugs	Tapones del oído	
Goggles	Anteojos	
Hard hat	Casco	
Lanyard	Acollador	Hard hat Casco
Perimeter guard	Guardado de proximidad	
Respirator	Respirador	
Fall harness	Caída de arnes	
Warning tape	Cinta de cuidado	Warning tape Cinta de cuidado
Vests	Chalecos	
Rebar caps	Tapadera de barilla	
Circuit breaker	Interruptor de circuito	
Rubber boots	Botas de goma	
Rubber gloves	Guantes de goma	
Respirator	Respirador	
Fire extinguisher	Extintor	
Restraint	Sujetador	Fire extinguisher Extintor

ACTIONS — ACCIONES	
English	*Español*
Bleed	Sangrar
Burn	Quemar
Call	Llamar
Evacuate	Evacuar
Fall	Caerse
Shock	Sacudida eléctrica
Slip	Deslizare
Stay off	Quedarse fuera
Rescue	Salvar

DESCRIPTIONS — DESCRIPCIONES	
Caution	Precaución
Combustible	Combustible
Secured	Asegurado, seguro
Caustic	Cáustico
Explosive	Explosivo
Flamable	Inflamable
Hazardous	Peligroso

PHRASES — FRASES

English	Español
Go for help.	Vaya por ayuda.
Call for help.	Llama ayuda.
Are you hurt?	¿Está herido?
Don't move.	No se mueva.
Do you use medicine?	¿Usa usted alguna medicina?
Do you use alcohol?	¿Usa usted alcohol?
Do you use drugs?	¿Usa usted drogas nórcoticos?
Get the first aid kit.	Traigame el juego de primeros auxilios.
Be careful.	Tenga cuidado.
Watch out.	Poner atención.
Authorized personnel only.	Personal autorizado solamente.
No smoking.	No fumar.
Keep out.	Manténgase fuera.
Where does it hurt?	¿Dónde le duele?
Pay attention to all warning signs.	Presta atención a todos las señales de peligro.
Report all accidents to your supervisor immediately.	Reporte todos los acidentes a su supervisor inmediamente.
Hard hats and safety glass are required at all times.	El casco y las gafas de seguridad se requieren siempre.

PHRASES (cont.) — FRASES (cont.)

English	Español
The first aid kit is in the trailer.	El juego de primeros auxilios está situado en el trailer.
Keep off the roof unless you are tied off.	No se suba al techo al menos que esté atado.
Where is the water jug?	¿Dónde está la jarra de agua potable?
Stay clear.	Quedarse alejado.

CHAPTER 14/CAPITULO 14
Tools and Equipment
Herramientas y Equipo

HAND TOOLS — HERRAMIENTAS DE MANO		
English	*Español*	
Hammer	Martillo	
Claw hammer	Martillo chivo	
Saw	Serrucho	Hammer Martillo
Hand saw	Serrucho de mano	
Bit and brace	Taladro de mano	
Drill	Taladro	
Drill bit	Broca, mecha	
Knife	Cuchillo	
Utility knife	Navaja de utilidad	Saw Serrucho
Pliers	Alicates, pinzas	
Channel-lock pliers	Alicates de extensión	
Vise-grip pliers	Alicates de presión, pinzas perras	
Screwdriver	Destornillador, desarmador	 Pliers Alicates, pinzas
Philips screwdriver	Desarmador cruz	
		 Philips screwdriver Desarmador cruz

HAND TOOLS *(cont.)* — HERRAMIENTAS DE MANO *(cont.)*

English	*Español*	
Sheet metal shears	Tijeras para metal	
Adjustable wrench	Llave francesa	
Crescent wrench	Llave de tuercas, llave adjustable	Adjustable wrench Llave francesa
Hook	Gancho	
Level	Nivel	
		Level Nivel

TOOLS WITH HANDLES — HERRAMIENTAS CON ASAS

English	Español	
Axe	Hacha	
Bar	Barreta	
Claw hammer	Martillo chivo	Axe Hacha
Mallet	Mazo	
Pick axe	Zapapico	
Shovel	Pala	
Sledgehammer	Marro, mazo	Shovel Pala
Hoe	Azadón	
Rake	Rastrillo	
Handle	Manija	
		Rake Rastrillo

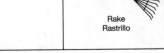

WOODWORKING TOOLS —
HERRAMIENTAS PARA MADERA

English	Español	
Hammer	Martillo	
Claw hammer	Martillo chivo	
Nail hammer	Martillo de clavos	
Cat's paw	Pata de gato	Hammer Martillo
Nails	Clavos	
Finishing nails	Clavos sin cabeza	
Nail gun	Clavador neumático	
Nail set	Botador de clavos	
Glue, adhesive	Adhesiva	
Wood chisel	Escoplo, formón, cincel	Nails Clavos
Clamp	Grapa, abrazadera	
C clamp	Grapa en C	
Bit and brace	Taladro de mano	
Carpenter's apron	Mandil, delantal	
Carpenter's square	Escuadra	
Framing square	Escuadra	C clamp Grapa en C
Square	Escuadra	
Tape measure	Cinta de medir	
Torch	Antorcha	
Hand saw	Serrucho de mano	
Cut-off saw	Sierra para cortar	
Jigsaw	Sierra de vaiven	Tape measure Cinta de medir

WOODWORKING TOOLS (cont.) — HERRAMIENTAS PARA MADERA (cont.)

English	Español	
Table saw	Serrucho de mesa	Table saw Serrucho de mesa
Chain saw	Sierra de cadena	
Plane	Cepillo	
Pliers	Alicates, pinzas	
Router	Fresadora, contorneador	
Sander	Lijadora	Sander Lijadora
Jointer	Cepillo automático	
Jointer plane	Cepillo de mano	
Darby	Plana, flatacho	
Drill	Taladro	
Drill bit	Broca, mecha	
File	Lima	
Level	Nivel	
Mallet	Mazo	Drill bit Broca, mecha
Mask	Máscara, careta	
Mitre box	Caja de corte a angulos	
Mitre saw	Sierra de retroceso para ingletes	
Punches	Punzones	
Vise	Morsa	
Pry bar	Barra de palanca	Vise Morsa
Block	Bloque	

WOODWORKING TOOLS (cont.) — HERRAMIENTAS PARA MADERA (cont.)

English	Español	
Bolt	Perno	
Wrench	Llave inglesa	
Torque wrench	Llave dinamometrica	
Spud wrench	Llave de cola	
Powder nailer	Pístola de cartuchos para fíjación	Saw horse Banqueta de aserrado
Saw horse	Banqueta de aserrado	
Screwdriver	Desarmador	
Screws	Tornillos	Screwdriver Desarmador
Screw gun	Pistola descarmadora	
Plumb bob	Plomo	
String line	Linea de hilo	Plumb bob Plomo

POWER TOOLS — HERRAMIENTAS ELÉCTRICAS

English	Español	
Circular saw	Sierra circular de mano	
Worm-drive saw	Sierra circular con tornillo sinfin	
Circular saw blade	Disco	
Mitre saw	Sierra de retroceso para ingletes	Circular saw Sierra circular de mano
Radial arm saw	Serrucho guillotina	
Reciprocating saw	Sierra alternativa	
Chop saw	Serrucho tajadero	
Table saw	Serrucho de mesa	
Saw guard	Protector de serrucho eléctrico	Reciprocating saw Sierra alternativa
Electric drill	Taladro eléctrico	
Hammer drill	Rotamarillo	
Right angle drill	Taladro de ángulo recto	
Drill bit	Broca, mecha	
Screw gun	Pistola descarmadora	
Nail gun	Clavador neumático	
Router	Fresadora, contorneador	Sander Lijadora
Sander	Lijadora	
Stapling gun	Engrapadora automática	
Extension cord	Cable de extensión	
Jackhammer	Martillo neumático	
Chain saw	Sierra de cadena	Chain saw Sierra de cadena
Impact wrench	Llave de impacto	

14-6

SPECIALTY TOOLS — HERRAMIENTAS ESPECIALES

English	Español	
Basin wrench	Llave pico de gansa	
Chain wrench	Llave de cadena	
Pipe wrench	Llave de tubo	
Chain saw	Sierra de cadena	**Pipe wrench** / **Llave de tubo**
Come-along	Mordaza tiradora de alambre	
Winch	Torno	
Shovel	Pala	
Slag hammer	Martillo de escoria	
Rebar bender	Doblador de varilla	**Shovel** / **Pala**
Sledgehammer	Marro, mazo	
Soldering torch	Soplete	
Stapler	Engrapadora	
Stapling gun	Engrapadora automática	
Strap wrench	Llave de correa, llave de cincho	
Square trowel	Llana	**Sledgehammer** / **Marro, mazo**
Mason's trowel	Paleta de albañil	
Joint-filler trowel	Paleta de relleno	
Valve-seat wrench	Llave de asiento de válvula	
Snake	Serpiente	
Vise	Morsa	**Vise** / **Morsa**
Vice bench	Torno de banco	

SPECIALTY TOOLS (cont.) — HERRAMIENTAS ESPECIALES (cont.)

English	Español	
Torch	Antorcha	
Propane	Propano	
Pry bar	Barra de palanca	
Fish tape	Cinta pescadora	
Tracer	Alambre testigo	
Laser	Láser	
Chalk line	Llinea de tiza	
Metal scribe	Trazador de metal	
Stud finder	Buscador be montantes	
Funnel	Embudo	
File	Lima	
Vacuum	Aspirar, aspiradora	
Air compressor	Compressor de aire	
Grinder	Afilador	
Wire brush	Cepillo de alambre	
C clamp	Grapa en C	
Tweezers	Pinzas pequeñas	
Powder nailer	Pístola de cartuchos para fíjación	
Cat's paw	Pata de chiva gato	
Walkie talkie	Transceptor portátil	

Torch
Antorcha

File
Lima

Wire brush
Cepillo de alambre

Walkie talkie
Transceptor portátil

SPECIALTY TOOLS (cont.) — HERRAMIENTAS ESPECIALES (cont.)

English	Español	
Bucket	Cubeta	
Chain	Cadena	
Tripod	Tripode	
Tamper	Pisón	
Caulking gun	Pistola de calafeto	Bucket Cubeta
Paint brush	Brocha para pintar	
Putty knife	Espátula de masilla	
Paint roller	Rodillo para pintura	
Flashlight	Linterna	

Bucket
Cubeta

Caulking gun
Pistola de calafeto

Flashlight
Linterna

EQUIPMENT — EQUIPO

English	Español	
Jack	Gato	
Crane	Grúa	
Crane operator	Operador de grúa	
Sling	Eslinga	Crane Grúa
Winch	Torno	
Come-along	Mordaza tiradora de alambre	
Conveyor	Transportador mecánico	
Scissor lift	Plataforma hidráulica	Scissor lift Plataforma hidráulica
Scaffold	Andamio	
Scaffolding	Andamiaje	
Step ladder	Escalera de tijera	
Extension ladder	Escalera de extensión	Scaffolding Andamiaje
Extension cord	Cable de extensión	
		Extension ladder Escalera de extensión

EQUIPMENT (cont.) — EQUIPO (cont.)

English	Español	
Truck	Camión	
Pick-up truck	Camioneta	Pick-up truck / Camioneta
Dump truck	Volquete	
Flat bed truck	Camión de tarima	
Truck bed	Caja de camión	
Clutch	Embrague	
Truck tailgate	Puerta trasera de camion	
Reflectors	Reflectores	
Wheel barrow	Carretilla	Wheel barrow / Carretilla
Tarp	Lona	
Tie	Amarra, ligadura	
Gloves	Guantes	
Eye protection	Protección de ojos	
Ear plugs	Tapones del oído	
Dust mask	Mascarilla para polvo	Gloves / Guantes
Respirator	Respirador	
Leather gloves	Guantes de cuero	
Steel toe boots	Botas protectoras de la punta del pie	Steel toe boots / Botas prectores de la punta del pie

EQUIPMENT (cont.) — EQUIPO (cont.)

English	Español	
Welding apron	Delantal de soldadura	
Protection	Protección	
Fire extinguisher	Extintor	
Vests	Chalecos	
Rubber boots	Botas de goma	
Rubber gloves	Guantes de goma	
Generator	Generador	
Machine	Máquina	
Work light	Lámpara de trabajo	
Power strip	Zapatilla eléctrica	
Power supply	Fuente de electricidad	
Cart	Carreta	
Chain	Cadena	
Hose	Manguera	
Air compressor	Compressor de aíre	
Blower	Sopladora	
Job box	Caja de trabajo	

Fire extinguisher
Extintor

Power strip
Zapatilla eléctrica

Generator
Generador

Hose
Manguera

OTHER MATERIALS — OTROS MATERIALES

English	Español	
Screw	Tornillo	
Self-tapping screws	Tornillos autorroscantes	
Self-drilling screws	Tornillos autoperforantes	Screw / Tornillo
Fasteners	Anclajes sujetadores	
Bolt	Perno	
Nut	Tuerca	
Expansion bolt	Perno de expansión	
Stainless steel	Acero inoxidable	
Plastic	Plástico	
Spring	Resorte	Bolt / Perno
Strap	Fleje (cinta)	
Strapping	Flejes (cintas)	
Tarp	Lona	
Tin	Lata, chapa, estaño	
Wedge	Cuña	
Chicken wire	Alambre de pollo	Wedge / Cuña
Wire mesh	Tela metálica	
Bags	Bolsas	
Cords	Cuerdas	
Tag	Etiqueta	
Pencil	Lápiz	
Glue	Pegamento	Pencil / Lápiz
Grip	Agarre	

OTHER MATERIALS (cont.)
— OTROS MATERIALES (cont.)

English	Español	
Grommet	Arandela	
Handle	Manipular	
Meter	Medidor	
Portable	Portátil	
Rivet	Remanche	
Gauge (thickness)	Calbrador	
Strut	Puntal	

Handle
Manipular

Meter
Medidor

CHAPTER 15/CAPITULO 15
The Construction Business
El Negocio De Construcción

PEOPLE — PERSONAS		
English	*Español*	
Architect	Arquitecto	
Engineer	Ingeniero	
Owner	Dueño	
Contractor	Contratista	
Inspector	Inspector	
Building inspector	Inspector de obras	Architect Arquitecto
Estimator	Estimador	
Supervisor	Supervisor, supervisora	
Manager	Gerente	
Foreman	Capataz	
Journeyman	Oficial	
Apprentice	Aprendiz	Foreman Capataz
Carpenter	Carpintero	
Worker	Trabajador	
Technician	Técnico	
Helper	Ayudante	
		Worker Trabajador

DOCUMENTS — DOCUMENTOS

English	Español	
Document	Documentar, documento	
Drawings	Dibujos	
Architectural plans	Planos arquitectónicos	
Site plans	Planos de sitio	Drawings Dibujos
Structural plans	Planos estructurales	
Schedule	Horario	
Construction schedule	Cronograma de construcción	
Estimate	Estimación	
Bid	Oferta	
Contract	Contrato	
Notes	Notas	
Notice	Aviso	
Permit	Permiso	Contract Contrato
Certificate of occupancy	Certificado de uso	
Lien	Derecho de retención	
Bond	Bono	
Performance bond	Bono de ejecución	
Report	Reporte	
Job report	Reporte de trabajo	
Daily report	Reporte del día	
Notice to proceed	Aviso de proceder	
Change order	Cambiar el orden	Change order Cambiar el orden
Safety policy	Poliza de seguridad	

BUSINESS WORDS — PALABRAS DE NEGOCIO

English	Español
Agree	Convenir, arprobar
Approve	Aprobar
Disagree	Discrepar/Desaprobar
Communicate	Comunicar
Coordinate	Coordinar
Approved	Aprobado
Alternate, alternative	Suplente, alternativa
Allowance	Complemento
Scope	Alcance
Price	Precio
Unit price	Precio de modulo
Estimate	Estimación
Take off	Extracción de los planos
To calculate	Calcular
Modification	Modificación
Responsible	Responsable
Right	Derecho
Delay	Retrasar
Performance	Desempeño
Determine	Determinar
Test	Probar
Building department	Departamento de construcción

BUSINESS WORDS (cont.) — PALABRAS DE NEGOCIO (cont.)

English	Español
Fire Code	Código de Incendios
Insurance	Seguro
Premises	Local, sitio
Property	Propiedad
Private	Privado
Proportion	Proporción, dimensionar
Proportioned	Dimensionado, Proporsionado
Provision	Disposición
Release	Descarga, liberación
Rental	Alquiler

OTHER WORDS — OTRAS PALABRAS

English	Español
Occupant load	Número de ocupantes
Gross area	Área total
Means of egress	Medios de salida
Shell	Cascaron, cubierta
Tenant	Inquilino
Value	Valor
Valuation	Valuación
Enforce	Hacer cumplir
Lay out	Croquis
Retrofit	Retroajuste
Hazardous communications	Comunicaciones peligrosas
Public safety	Seguridad pública, protección al público
Essential facilities	Instalaciones esenciales
Sanitation	Higiene
Quota	Cuota
Rate	Relación, razón
Rating	Clasifación
Ratio	Relación, cociente
Region	Región, tramo
Shop	Taller
Repair	Reparación
Link, linkage	Enlace, tirante

NOTES — NOTAS

CHAPTER 16/CAPITULO 16
Housekeeping
Que Haceres

LOCATIONS — LOCACIONES		
English	*Español*	
Ceiling	Cielo	
Classroom	Sala de clase	Classroom / Sala de clase
Corridor	Pasillo	
Door	Puerta	
Doorway	Umbral	Door / Puerta
Entry	Entrada	
Exit	Salida	
Floor	Piso	
Gymnasium	Gimnasio	
Hallway	Vestíbulo	
Kitchen	Cocina	Gymnasium / Gimnasio
Washroom	Baño	
Wall	Pared	
Window	Ventana	Window / Ventana

EQUIPMENT — EQUIPO

English	Español	
Broom	Escoba	
Brush	Brocha, Cepillo	
Bucket	Cubeta	
Cart	Carreta	
Ceiling	Cielo	
Chemical	Producto, quimico	
Door	Puerta	
Drawer	Cajone	
Dumpster	Tambo de basura	
Dust mop	Trapeador de polvo	
Dust pan	Recogedor de polvo	

Broom
Escoba

Bucket
Cubeta

Door
Puerta

Dumpster
Tarrbo de basura

EQUIPMENT (cont.) — EQUIPO (cont.)

English	Español	
Faucet	Llave de agua	
Floor	Piso	
Garden hose	Manguera de jardín	
Hand brush	Cepillo de mano	Faucet / Llave de agua
Mop	Trapeador	
Mop bucket	Cubeta para trapeador	
Pile	Pila	
Paper dispensers	Dispensadores de papel	
Push broom	Escobón	
Rags	Trapos	Mop / Trapeador
Sink	Fregadero	
Soap	Jabón	
Spray bottle	Rociador	
Squeegee	Enjugador de vidrio	
Trash	Basura	Sink / Fregadero
Vacuum	Aspiradora	
Wash basin	Lavabo	
Warm water	Agua tibia	
Cold water	Agua fría	
Hot water	Agua caliente	Vacuum / Aspiradora

ACTIONS — ACCIONES

English	Español
Add	Agregar
Apply	Aplicar
Check	Comprobar
Communicate	Comunicar
Coordinate	Coordinar
Divide	Dividir
Brush	Cepillar
Change	Cambiar
Clean	Limpiar
Drain	Drenar
Empty	Vaciar
Dust	Quitar el polvo
Lift	Elevar
Mop	Trapear
Pick up	Recoger
Polish	Pulir
Rinse	Enjuagar
Scrape	Raspar
Scrub	Fregar

ACTIONS (cont.) — ACCIONES (cont.)

English	Español
Shine	Brillar
Spray	Rociar
Store	Almacenar
Sweep	Barrer
Vacuum	Pasar la aspiradora
Wait	Esperar
Walk	Caminar
Wash	Lavar
Wipe	Frotar ligeramente

WEATHER — TIEMPO

English	Español
Cold	Frío
Hot	Caliente
Freezing	Congelante
Frost	Helada
Rainy	Lluvioso
Slick	Liso
Slippery	Resbaladizo
Snow	Nieve
Windy	Ventoso
Wet	Mojado
Muddy	Fangoso (tierra)
Slushy	Fangoso (nieve)

DESCRIPTIONS — DESCRIPCIONES

English	Español
Damp	Húmedo
Dirty	Sucío
Dry	Seco
Clean	Limpio
Inside	Adentro
Outside	Afuera
Shiny	Brillante
Wet	Mojado
On time	A tiempo
Not ready yet.	No está listo todavía.
Precisely (time)	En punto (tiempo)

PHRASES — FRASES

English	Español
You missed a spot.	Faltó un punto.
Put up the "Slippery When Wet" sign when you mop.	Ponga la señal "Resbaloso Cuando Mojado" cuando usted trapee."
Change the water in your bucket often.	Cambia el agua de la cubeta a menudo.
The floor has been waxed, you'll have to wait until it is dry to walk on it.	Se he encerado el piso, tendra usted que esperar hasta que se seque antes de caminar en él.
Use sweeping compound when you sweep the floor.	Utilice el compuesto de barrida cuando barra el piso.

CHAPTER 17/CAPITULO 17
Basic Words and Phrases
Palabras y Frases Básicas

BASIC WORDS — PALABRAS BÁSICAS	
English	*Español*
Who	Quién
What	Qué
When	Cuándo
Where	Donde
How	Como
Why	Porqué
This	Este
That	Ese, esa, eso
Of	De
For	Por, para
But	Pero
Correct, right	Correcto
Incorrect, wrong	Incorrecto
Better	Mejor
Change	Cambio
Want	Quiero
Hello	Hola
Goodbye	Adiós
And	Y

BASIC WORDS (cont.) — PALABRAS BÁSICAS (cont.)

English	Español
Or	O
With	Con
Because	Por que
For	Por
In	En
To	A
From	De
Yes	Sí
No	No
Off	Fuera de, apagado
On	En, sobre
Once	Una vez
Again	Otra vez
Only	Solamente
Open	Abra, abrir
Go (you)	Ve, ir
Stop	Parar
Problem	Problema
Same	Uniforme
I'm sorry	Lo siento

BASIC WORDS (cont.) — PALABRAS BÁSICAS (cont.)

English	Español
Alert	Alerta, vigilante
Gap	Brecha
Always	Siempre
Never	Nunca
Here	Aquí
There	Allá
OK	OK
Old	Viejo
Pen	Pluma
Point (shape)	Punta
Problem	Problema

PEOPLE — PERSONAS

English	Español
Family	Familia
Man	Hombre
Woman	Señorita, Señora
Boss	Jefe
Friend	Amigo
Us	Nosotros
Them	Ellos
Supervisor	Supervisor
I	Yo
You	Tu, usted
He	Él
She	Ella
We	Nosotros
Them	Ellos
My	Mi
Your	Tus

ACTIONS — ACCIONES

English	Español
To give	Dar
To take	Tomar
To lift	Levantar
To release	Soltar
To put	Poner, colocar
To pour	Echar
To have	Tener
To come	Venir
To go	Ir
To build	Construir
To move	Mueva
To work	Trabajo
To repeat	Repite
To find	Encontrar
To know	Saber
To look	Mirar
To stop	Parar
To start	Empezar
To speak	Hablar
To fix	Arregle
To need	Necesitar

DESCRIPTIONS — DESCRIPCIONES

English	Español
Much	Mucho
Little	Poquito
More	Más
Less	Menos
Hard	Duro, firme
Soft	Blando, suave
Wet	Mojado
Dry	Seco
Loud	Alto, Ruidoso
Quiet	Tranquilo, silencío
Safe	Ileso, seguro
Dangerous	Peligroso
Fast	Rápido
Slow	Despacío
Enough, sufficient	Suficiente
Under	Debajo
Over	Encima
Good	Bien
Bad	Malo
Large	Grande
Small	Pequeño

DESCRIPTIONS (cont.) — DESCRIPCIONES (cont.)

English	Español
Same	Mismo, igual
Different	Diferente
Hot	Caliente
Cold	Frío
Many	Muchos
Few	Pocos
First	Primera
Last	Última
Easy	Fácil
Difficult	Difícil
Near	Cerca
Far	Lejos
Front	Frente, delante
A few	Pocos
Middle	Medio
Near	Cerca, proximo
Quickly	Pronto
Corrosion resistant	Anticorrosivo
Corrosive	Corrosivo

TIME — TIEMPO	
English	*Español*
Hour	Hora
Minute	Minuto
Today	Hoy
Tomorrow	Mañana
Morning	La mañana
Afternoon	Tarde
Night, evening	Noche
Before	Antes
After	Después
On time	A tiempo
Late	Tarde
Now	Ahora
Later	Más tarde
Always	Siempre
Never	Nunca
Not ready yet.	No está listo todavía.
Precisely (time)	En punto (tiempo)

DAYS — DIAS	
English	*Español*
Monday	Lunes
Tuesday	Martes
Wednesday	Miércoles
Thursday	Jueves
Friday	Viernes
Saturday	Sábado
Sunday	Domingo
Holiday	Día feriado
MONTHS — MESES	
January	Enero
February	Febrero
March	Marzo
April	Abril
May	Mayo
June	Junio
July	Julio
August	Agosto
September	Septiembre
October	Octubre
November	Noviembre
December	Diciembre

NUMBERS — NUMEROS

	English	*Español*
0	Zero	Cero
1	One	Uno
2	Two	Dos
3	Three	Tres
4	Four	Cuatro
5	Five	Cinco
6	Six	Seis
7	Seven	Siete
8	Eight	Ocho
9	Nine	Nueve
10	Ten	Diez
11	Eleven	Once
12	Twelve	Doce
13	Thirteen	Trece
14	Fourteen	Catorce
15	Fifteen	Quince
16	Sixteen	Dieciséis
17	Seventeen	Diecisiete
18	Eighteen	Dieciocho
19	Nineteen	Diecinueve
20	Twenty	Veinte
21	Twenty-one	Veintiuno
22	Twenty-two	Veintidós

NUMBERS (cont.) — NUMEROS (cont.)

	English	Español
30	Thirty	Treinta
31	Thirty-one	Treinta y uno
40	Forty	Cuarenta
50	Fifty	Cincuenta
60	Sixty	Sesenta
70	Seventy	Setenta
80	Eighty	Ochenta
90	Ninety	Noventa
100	One hundred	Cien
101	One hundred and one	Ciento uno
200	Two hundred	Doscientos
300	Three hundred	Trescientos
400	Four hundred	Cuatrocientos
500	Five hundred	Quinientos
600	Six hundred	Seiscientos
700	Seven hundred	Setecientos
800	Eight hundred	Ochocientos
900	Nine hundred	Novecientos
1,000	One thousand	Mil
2,000	Two thousand	Dos mil
1,000,000	One million	Un millón
2,000,000	Two million	Dos millones

PHRASES — FRASES

English	Español
Give me	Da me
What is?	¿Qué es?
How much?	¿Cuánto?
How many?	¿Cuántos?
What time?	¿Qué hora?
I need	Necesito, necesitar
I don't understand.	No comprendo.
What does____ mean?	¿Que es lo que significa ___?
How do you say___?	¿Cómo se dice ___?
Do you speak English?	Hablas Inglés?
Do you speak Spanish?	Hablas Español?
Work safely.	Trabaje con cuidado.
Move this.	Mueva este. (esto)
Get me the ___.	Tráigame el (la) ___.
Clean this.	Limpie esto.
Good work.	Buen trabajo.
Where is your ___?	¿Dónde está su ___?
Can someone interpret?	¿Puede alguien interpretar?
Keep the job site clean.	Mantenga el trabajo limpio.
You're welcome.	Gracias.
I'm sorry.	Lo siento.
Excuse me. (permission)	Con permiso.
Excuse me. (apology)	Disculpe.

CHAPTER 18/CAPITULO 18
Measurement and Conversion Factors
Unidades de Medida y Factor de Conversiones

In this chapter you will find conversion factors and units of measurement. These are shown with all-English on the left-hand page and all-Spanish on the right page. The two pages will almost always be identical. So, the fifth line from the bottom on the left will convey exactly the same information as the fifth line from the bottom on the right.

En esté capítulo usted encontrará factores de la conversión y unidades de medida. Éstos se demuestran con solo-Inglés en el lado izquierdo de la página y solo-Español en la página derecha. Las dos páginas serán siempre idénticas. Así pues, la quinta línea del fondo a la izquierda transportará exactamente la misma información que la quinta línea del fondo a la derecha.

COMMONLY USED CONVERSION FACTORS

Multiply	By	To Obtain
Acres	43,560	Square feet
Acres	1.562×10^{-3}	Square miles
Acre-Feet	43,560	Cubic feet
Amperes per sq cm	6.452	Amperes per sq in.
Amperes per sq in.	0.1550	Amperes per sq cm
Ampere-Turns	1.257	Gilberts
Ampere-Turns per cm	2.540	Ampere-turns per in.
Ampere-Turns per in.	0.3937	Ampere-turns per cm
Atmospheres	76.0	Cm of mercury
Atmospheres	29.92	Inches of mercury
Atmospheres	33.90	Feet of water
Atmospheres	14.70	Pounds per sq in.
British thermal units	252.0	Calories
British thermal units	778.2	Foot-pounds
British thermal units	3.960×10^{-4}	Horsepower-hours
British thermal units	0.2520	Kilogram-calories
British thermal units	107.6	Kilogram-meters
British thermal units	2.931×10^{-4}	Kilowatt-hours
British thermal units	1,055	Watt-seconds
B.t.u. per hour	2.931×10^{-4}	Kilowatts
B.t.u. per minute	2.359×10^{-2}	Horsepower
B.t.u. per minute	1.759×10^{-2}	Kilowatts
Bushels	1.244	Cubic feet
Centimeters	0.3937	Inches
Circular mils	5.067×10^{-6}	Square centimeters
Circular mils	0.7854×10^{-6}	Square inches
Circular mils	0.7854	Square mils
Cords	128	Cubic feet
Cubic centimeters	6.102×10^{-6}	Cubic inches
Cubic feet	0.02832	Cubic meters
Cubic feet	7.481	Gallons
Cubic feet	28.32	Liters
Cubic inches	16.39	Cubic centimeters
Cubic meters	35.31	Cubic feet

FACTORES DE CONVERSIÓN DE USO COMÚN

Multiplique	Por	Para obtener
Acres	43,560	Pies cuadrados
Acres	1.562×10^{-3}	Millas cuadradas
Acre-pies	43,560	Pies cúbicos
Amperes por centímetro cuadrado	6.452	Amperes por pulgada cuadrada
Amperes por pulgada cuadrada	0.1550	Amperes por centímetro cuadrado
Amperes-vueltas	1.257	Gilberts
Amperes-vueltas por cm.	2.540	Amperes-vueltas por pulgada
Amperes-vueltas por pulgada	0.3937	Amperes-vueltas por cm
Atmósferas	76.0	Centímetros de mercurio
Atmósferas	29.92	Pulgadas de mercurio
Atmósferas	33.90	Pies de agua
Atmósferas	14.70	Libras por pulgada cuadrada
Brazas	6	Pies
B.t.u. por hora	2.931×10^{-4}	Kilowatts
B.t.u. por minuto	2.359×10^{-2}	Potencia en HP
B.t.u. por minuto	1.759×10^{-2}	Kilowatts
Bushels	1.244	Pies cúbicos
Centímetros	0.3937	Pulgadas
Centímetros cuadrados.	1.973×10^{5}	Milímetros circulares
Centímetros cúbicos	6.102×10^{-6}	Pulgadas cúbicas
Cuerdas	128	Pies cúbicos
Dinas	2.248×10^{-6}	Libras
Ergs	1	Dinas-centímetros
Ergs	7.37×10^{-6}	Libras-pie
Ergs	10^{-7}	Joules
Farads	10^{6}	Microfarads
Galones	0.1337	Pies cúbicos
Galones	231	Pulgadas cúbicas
Galones	3.785×10^{-3}	Metros cúbicos
Galones	3.785	Litros
Galones por minuto	2.228×10^{-3}	Pies cúbicos por segundo
Gausses	6.452	Líneas por pulgada cuadrada
Gilberts	0.7958	Amperes-vueltas
Grados (ángulo)	0.01745	Radianes

COMMONLY USED CONVERSION FACTORS (cont.)

Multiply	By	To Obtain
Cubic meters	1.308	Cubic yards
Cubic yards	0.7646	Cubic meters
Degrees (angle)	0.01745	Radians
Dynes	2.248×10^{-6}	Pounds
Ergs	1	Dyne-centimeters
Ergs	7.37×10^{-6}	Foot-pounds
Ergs	10^{-7}	Joules
Farads	10^6	Microfarads
Fathoms	6	Feet
Feet	30.48	Centimeters
Feet of water	.08826	Inches of mercury
Feet of water	304.8	Kg per square meter
Feet of water	62.43	Pounds per square ft.
Feet of water	0.4335	Pounds per square in.
Foot-pounds	1.285×10^{-2}	British thermal units
Foot-pounds	5.050×10^{-7}	Horsepower-hours
Foot-pounds	1.356	Joules
Foot-pounds	0.1383	Kilogram-meters
Foot-pounds	3.766×10^{-7}	Kilowatt-hours
Gallons	0.1337	Cubic feet
Gallons	231	Cubic inches
Gallons	3.785×10^{-3}	Cubic meters
Gallons	3.785	Liters
Gallons per minute	2.228×10^{-3}	Cubic feet per sec.
Gausses	6.452	Lines per square in.
Gilberts	0.7958	Ampere-turns
Henries	10^3	Millihenries
Horsepower	42.41	Btu per min.
Horsepower	2,544	Btu per hour
Horsepower	550	Foot-pounds per sec.
Horsepower	33,000	Foot-pounds per min.
Horsepower	1.014	Horsepower (metric)
Horsepower	10.70	Kg calories per min.
Horsepower	0.7457	Kilowatts

FACTORES DE CONVERSIÓN DE USO COMÚN (cont.)

Multiplique	Por	Para obtener
Henrys	10^3	Milihenrys
Joules	9.478×10^{-4}	Unidades térmicas del sistema inglés
Joules	0.2388	Calorías
Joules	10^7	Ergs
Joules	0.7376	Libras-pie
Joules	2.778×10^{-7}	Kilowatt-horas
Joules	0.1020	Kilogramo-metros
Joules	1	Watts-segundo
Kg por metro cuadrado	3.281×10^{-3}	Pies de agua
Kg por metro cuadrado	0.2048	Libras por pie cuadrado
Kg por metro cuadrado	1.422×10^{-3}	Libras por pulgada cuadrada
Kilogramos	2.205	Libras
Kilogramo-calorías	3.968	Unidades térmicas del sistema inglés
Kilogramo metros	7.233	Libras-pie
Kilolíneas	10^3	Maxwells
Kilómetros	3.281	Pies
Kilómetros	0.6214	Millas
Kilómetros cuadrados	0.3861	Millas cuadradas
Kilowatts	56.87	B.t.u. por minuto
Kilowatts	737.6	Libras-pie por segundo
Kilowatts	1.341	Potencia en HP
Kilowatts-horas	3409.5	Unidades térmicas del sistema inglés
Kilowatts-horas	2.655×10^6	Libras-pie
Libras	32.17	Poundals
Libras de agua	0.01602	Pies cúbicos
Libras de agua	0.1198	Galones
Libras-pie	1.285×10^{-2}	Unidades térmicas del sistema inglés
Libras-pie	5.050×10^{-7}	Potencia en HP-horas
Libras-pie	1.356	Joules
Libras-pie	0.1383	Kilogramo-metros
Libras-pie	3.766×10^{-7}	Kilowatt-horas'
Libras-pies	0.1383	Metro-kilogramos
Libras por pie cuadrado	6.944×10^{-3}	Libras por pulgada cuadrada
Libras por pie cuadrado	0.01602	Pies de agua

COMMONLY USED CONVERSION FACTORS (cont.)

Multiply	By	To Obtain
Horsepower (boiler)	33,520	Btu per hour
Horsepower-hours	2,544	British thermal units
Horsepower-hours	1.98×10^6	Foot-pounds
Horsepower-hours	2.737×10^5	Kilogram-meters
Horsepower-hours	0.7457	Kilowatt-hours
Inches	2.540	Centimeters
Inches of mercury	1.133	Feet of water
Inches of mercury	70.73	Pounds per square ft.
Inches of mercury	0.4912	Pounds per square in.
Inches of water	25.40	Kg per square meter
Inches of water	0.5781	Ounces per square in.
Inches of water	5.204	Pounds per square ft.
Joules	9.478×10^{-4}	British thermal units
Joules	0.2388	Calories
Joules	10^7	Ergs
Joules	0.7376	Foot-pounds
Joules	2.778×10^{-7}	Kilowatt-hours
Joules	0.1020	Kilogram-meters
Joules	1	Watt-seconds
Kilograms	2.205	Pounds
Kilogram-calories	3.968	British thermal units
Kilogram meters	7.233	Foot-pounds
Kg per square meter	3.281×10^{-3}	Feet of water
Kg per square meter	0.2048	Pounds per square ft.
Kg per square meter	1.422×10^{-3}	Pounds per square in.
Kilolines	10^3	Maxwells
Kilometers	3.281	Feet
Kilometers	0.6214	Miles
Kilowatts	56.87	Btu per min.
Kilowatts	737.6	Foot-pounds per sec.
Kilowatts	1.341	Horsepower
Kilowatts-hours	3409.5	British thermal units
Kilowatts-hours	2.655×10^6	Foot-pounds
Knots	1.152	Miles

FACTORES DE CONVERSIÓN DE USO COMÚN (cont.)

Multiplique	Por	Para obtener
Libras por pie cuadrado	4.882	Kg. por metro cuadrado
Libras por pulgada cuadrada .	2.307	Pies de agua
Libras por pulgada cuadrada .	2.036	Pulgadas de mercurio
Libras por pulgada cuadrada .	703.1	Kg. por metro cuadrado
Libras por pulgada cúbica	27.68	Gramos por centímetro cúbico
Libras por pulgada cúbica	2.768×10^{-4}	Kg. por metro cúbico
Libras por pulgada cúbica	1.728	Libras por pie cúbico
Libras por pie cúbico	16.02	Kg. por metro cúbico
Libras por pie cúbico	5.787×10^{-4}	Libras por pulgada cúbica
Litros	0.03531	Pies cúbicos
Litros	61.02	Pulgadas cúbicas
Litros	0.2642	Galones
Log N	2.303	Log_e N o en base N
Log N_e o en base N	0.4343	Log_{10} N
Lúmenes por pie cuadrado. . .	1	Candelas pie
Maxwells	10^{-3}	Kilolíneas
Megalíneas	10^{6}	Maxwells
Megohms	10^{6}	Ohms
Metros	3.281	Pies
Metros	39.37	Pulgadas
Metros cuadrados	10.76	Pies cuadrados
Metros cúbicos	35.31	Pies cúbicos
Metros cúbicos	1.308	Yardas cúbicas
Metro-kilogramos	7.233	Libras-pies
Microfarads	10^{-6}	Farads
Microhms	10^{-6}	Ohms
Microhms por cm cúbico	0.3937	Microhms por pulgada cúbica
Microhms por cm cúbico	6.015	Ohms por pies-mil
Millas	5,280	Pies
Millas	1.609	Kilómetros
Milímetros circulares	5.067×10^{-6}	Centímetros cuadrados
Milímetros circulares	0.7854×10^{-6}	Pulgadas cuadradas
Milímetros circulares	0.7854	Mils cuadrados
Millas cuadradas	640	Acres
Millas cuadradas	2.590	Kilómetros cuadrados
Milímetros cuadrados	1.973×10^{3}	Milímetros circulares
Mils cuadrados	1.273	Milímetros circulares

COMMONLY USED CONVERSION FACTORS (cont.)

Multiply	By	To Obtain
Liters	0.03531	Cubic feet
Liters	61.02	Cubic inches
Liters	0.2642	Gallons
Log N_e or in N	0.4343	Log_{10} N
Log N	2.303	Log_e N or in N
Lumens per square ft.	1	Footcandles
Maxwells	10^{-3}	Kilolines
Megalines	10^6	Maxwells
Megaohms	10^6	Ohms
Meters	3.281	Feet
Meters	39.37	Inches
Meter-kilograms	7.233	Pound-feet
Microfarads	10^{-6}	Farads
Microhms	10^{-6}	Ohms
Microhms per cm cube	0.3937	Microhms per in. cube
Microhms per cm cube	6.015	Ohms per mil foot
Miles	5,280	Feet
Miles	1.609	Kilometers
Miner's inches	1.5	Cubic feet per min.
Ohms	10^{-6}	Megohms
Ohms	10^6	Microhms
Ohms per mil foot	0.1662	Microhms per cm cube
Ohms per mil foot	0.06524	Microhms per in. cube
Poundals	0.03108	Pounds
Pounds	32.17	Poundals
Pound-feet	0.1383	Meter-Kilograms
Pounds of water	0.01602	Cubic feet
Pounds of water	0.1198	Gallons
Pounds per cubic foot	16.02	Kg per cubic meter
Pounds per cubic foot	5.787×10^{-4}	Pounds per cubic in.
Pounds per cubic inch	27.68	Grams per cubic cm
Pounds per cubic inch	2.768×10^{-4}	Kg per cubic meter
Pounds per cubic inch	1.728	Pounds per cubic ft.
Pounds per square foot	0.01602	Feet of water

FACTORES DE CONVERSIÓN DE USO COMÚN (cont.)

Multiplique	Por	Para obtener
Nudos	1.152	Millas
Ohms	10^{-6}	Megohms
Ohms	10^{6}	Microhms
Ohms por pies-mil	0.1662	Microhms por cm cúbico
Ohms por pies-mil	0.06524	Microhms por pulgada cúbica
Pies	30.48	Centímetros
Pies cuadrados	2.296×10^{-5}	Acres
Pies cuadrados	0.09290	Metros cuadrados
Pies cúbicos	0.02832	Metros cúbicos
Pies cúbicos	7.481	Galones
Pies cúbicos	28.32	Litros
Pies de agua	.08826	Pulgadas de mercurio
Pies de agua	304.8	Kg. por metro cuadrado
Pies de agua	62.43	Libras por pie cuadrado
Pies de agua	0.4335	Libras por pulgada cuadrada
Potencia en HP	42.41	B.t.u. por minuto
Potencia en HP	2,544	B.t.u. por hora
Potencia en HP	550	Libras-pie por segundo
Potencia en HP	33,000	Libras-pie por minuto
Potencia en HP	1.014	Potencia en HP (unidades métricas)
Potencia en HP	10.70	Kg. calorías por minuto
Potencia en HP	0.7457	Kilowatts
Potencia en HP (caldera)	33,520	B.t.u. por hora
Potencia en HP-horas	2,544	Unidades térmicas del sistema inglés
Potencia en HP-horas	1.98×10^{6}	Libras-pie
Potencia en HP-horas	2.737×10^{5}	Kilogramo-metros
Potencia en HP-horas	0.7457	Kilowatt-horas
Poundals	0.03108	Libras
Pulgadas	2.540	Centímetros
Pulgadas cuadradas	1.273×10^{6}	Milímetros circulares
Pulgadas cuadradas	6.452	Centímetros cuadrados
Pulgadas cúbicas	16.39	Centímetros cúbicos
Pulgadas de agua	25.40	Kg. por metro cuadrado
Pulgadas de agua	0.5781	Onzas por pulgada cuadrada
Pulgadas de agua	5.204	Libras por pie cuadrado
Pulgadas de mercurio	1.133	Pies de agua

Multiply	By	To Obtain
Pounds per square foot	4.882	Kg per square meter
Pounds per square foot	6.944×10^{-3}	Pounds per sq. in.
Pounds per square inch	2.307	Feet of water
Pounds per square inch	2.036	Inches of mercury
Pounds per square inch	703.1	Kg per square meter
Radians	57.30	Degrees
Square centimeters	1.973×10^{5}	Circular mils
Square feet	2.296×10^{-5}	Acres
Square feet	0.09290	Square meters
Square inches	1.273×10^{6}	Circular mils
Square inches	6.452	Square centimeters
Square kilometers	0.3861	Square miles
Square meters	10.76	Square feet
Square miles	640	Acres
Square miles	2.590	Square kilometers
Square millimeters	1.973×10^{3}	Circular mils
Square mils	1.273	Circular mils
Tons (long)	2,240	Pounds
Tons (metric)	2,205	Pounds
Tons (short)	2,000	Pounds
Watts	0.05686	Btu per minute
Watts	10^{7}	Ergs per sec.
Watts	44.26	Foot-pounds per min.
Watts	1.341×10^{-3}	Horsepower
Watts	14.34	Calories per min.
Watts-hours	3.412	British thermal units
Watts-hours	2,655	Footpounds
Watts-hours	1.341×10^{-3}	Horsepower-hours
Watts-hours	0.8605	Kilogram-calories
Watts-hours	376.1	Kilogram-meters
Webers	10^{8}	Maxwells

FACTORES DE CONVERSIÓN DE USO COMÚN (cont.)

Multiplique	Por	Para obtener
Pulgadas de mercurio	70.73	Libras por pie cuadrado
Pulgadas de mercurio	0.4912	Libras por pulgada cuadrada
Pulgadas mineras	1.5	Pies cúbicos por minuto
Radianes	57.30	Grados
Toneladas (largas)	2,240	Libras
Toneladas (métricas)	2,205	Libras
Toneladas (cortas)	2,000	Libras
Unidades térmicas del sistema inglés	252.0	Calorías
Unidades térmicas del sistema inglés	778.2	Libras-pie
Unidades térmicas del sistema inglés	3.960×10^{-4}	Potencia en HP-horas
Unidades térmicas del sistema inglés	0.2520	Kilogramo-calorías
Unidades térmicas del sistema inglés	107.6	Kilogramo-metros
Unidades térmicas del sistema inglés	2.931×10^{-4}	Kilowatt-horas
Unidades térmicas del sistema inglés	1,055	Watts-segundo
Watts	0.05686	B.t.u. por minuto
Watts	10^{7}	Ergs por segundo
Watts	44.26	Libras-pie por minuto
Watts	1.341×10^{-3}	Potencia en HP
Watts	14.34	Calorías por minuto
Watts-hora	3.412	Unidades térmicas del sistema inglés
Watts-hora	2,655	Libras-pie
Watts-hora	1.341×10^{-3}	Caballo de fuerzas-horas
Watts-hora	0.8605	Kilogramo-calorías
Watts-hora	376.1	Kilogramo-metros
Webers	10^{8}	Maxwells
Yardas cúbicas	0.7646	Metros cúbicos

ELECTRICAL PREFIXES

Prefixes
Prefixes are used to avoid long expressions of units that are smaller and larger than the base unit. See Common Prefixes. For example, sentences 1 and 2 do not use prefixes. Sentences 3 and 4 use prefixes.
1. A solid-state device draws 0.000001 amperes (A).
2. A generator produces 100,000 watts (W).
3. A solid-state device draws 1 microampere (μA).
4. A generator produces 100 kilowatts (kW).

Converting Units
To convert between different units, the decimal point is moved to the left or right, depending on the unit. See Conversion Table. For example, an electronic circuit has a current flow of .000001 A. The current value is converted to simplest terms by moving the decimal point six places to the right to obtain 1.0μA (from Conversion Table).

$$.000001 \, A = 1.0 \, \mu A$$

Move decimal point
6 places to right

Common Electrical Quantities
Abbreviations are used to simplify the expression of common electrical quantities.
See Common Electrical Quantities. For example, milliwatt is abbreviated mW, kilovolt is abbreviated kV, and ampere is abbreviated A.

COMMON PREFIXES

Symbol	Prefix	Equivalent
G	giga	1,000,000,000
M	mega	1,000,000
k	kilo	1000
base unit	—	1
m	milli	.001
u	micro	.000001
n	nano	.000000001

COMMON ELECTRICAL QUANTITIES

Variable	Name	Unit of Measure and Abbreviation
E	voltage	volt - V
I	current	ampere - A
R	resistance	ohm - Ω
P	power	watt - W
P	power (apparent)	volt-amp - VA
C	capacitance	farad - F
L	inductance	henry - H
Z	impedance	ohm - Ω
G	conductance	siemens - S
f	frequency	hertz - Hz
T	period	second - s

CONVERSION TABLE

Initial Units	Final Units						
	giga	mega	kilo	base unit	milli	micro	nano
giga	—	3R	6R	9R	12R	15R	18R
mega	3L	—	3R	6R	9R	12R	15R
kilo	6L	3L	—	3R	6R	9R	12R
base unit	9L	6L	3L	—	3R	6R	9R
milli	12L	9L	6L	3L	—	3R	6R
micro	15L	12L	9L	6L	3L	—	3R
nano	18L	15L	12L	9L	6L	3L	—

PREFIJOS ELÉCTRICOS

Prefijos

Los prefijos se usan para evitar el uso de largas expresiones para designar unidades que son más pequeñas o más grandes que la unidad base. Ver prefijos comunes. Por ejemplo, las frases 1 y 2 no usan prefijos. Las frases 3 y 4 usan prefijos.

1. Un dispositivo de estado sólido absorbe 0.000001 amperes (A).
2. Un generador produce 100,000 watts (W).
3. Un dispositivo de estado sólido absorbe 1 microamper (uA).
4. Un generador produce 100 kilowatts (kW).

Conversión de unidades

Para convertir entre unidades diferentes, el punto decimal se desplaza a la izquierda o la derecha, según sea la unidad. Ver Tabla de conversión. Por ejemplo, un circuito electrónico tiene un flujo de corriente de .000001 A. El valor de la corriente se convierte a los términos más simples desplazando el punto decimal seis lugares hacia la derecha para obtener 1.0 µA (de la Tabla de conversión).

$$.000001. \, A = 1.0 \, uA$$

Desplace el punto decimal
6 lugares a la derecha

Cantidades eléctricas comunes

Las abreviaturas se usan para simplificar la expresión de las cantidades eléctricas comunes. Ver cantidades eléctricas comunes. Por ejemplo, miliwatt se abrevia mW, kilovolt se abrevia kV y ampere se abrevia A

PREFIJOS COMUNES

Símbolo	Prefijo	Equivalencia
G	giga	1,000,000,000
M	mega	1,000,000
k	kilo	1000
unidad base	—	1
m	milli	.001
u	micro	.000001
n	nano	.000000001

CANTIDADES ELÉCTRICAS COMUNES

Variable	Nombre	Unidad de medida y abreviatura
E	voltaje	volt - V
I	corriente	ampere - A
R	resistencia	ohm - Ω
P	potencia	watt - W
P	potencia (aparente)	volt-amp - VA
C	capacidad	farad - F
L	inductancia	henry - H
Z	impedancia	ohm - Ω
G	conductancia	siemens - S
f	frecuencia	hertz - Hz
T	período	segundos - s

TABLA DE CONVERSIÓN

Unidades iniciales	Unidades finales						
	giga	mega	kilo	unidad base	mili	micro	nano
giga	—	3R	6R	9R	12R	15R	18R
mega	3L	—	3R	6R	9R	12R	15R
kilo	6L	3L	—	3R	6R	9R	12R
unidad base	9L	6L	3L	—	3R	6R	9R
milli	12L	9L	6L	3L	—	3R	6R
micro	15L	12L	9L	6L	3L	—	3R
nano	18L	15L	12L	9L	6L	3L	—

CONVERSION TABLE FOR TEMPERATURE – °F/°C

°F	°C	°F	°C	°F	°C	°F	°C	°F	°C
-459.4	-273	-22.0	-30	35.6	2	93.2	34	150.8	66
-418.0	-250	-18.4	-28	39.2	4	96	36	154.4	68
-328.0	-200	-14.8	-26	42.8	6	100.4	38	158.0	70
-238.0	-150	-11.2	-24	46.4	8	104.0	40	161.6	72
-193.0	-125	-7.6	-22	50.0	10	107.6	42	165.2	74
-148.0	-100	-4.0	-20	53.6	12	111.2	44	168.8	76
-130.0	-90	-0.4	-18	57.2	14	114.8	46	172.4	78
-112.0	-80	3.2	-16	60.8	16	118.4	48	176.0	80
-94.0	-70	6.8	-14	64.4	18	122.0	50	179.6	82
-76.0	-60	10.4	-12	68.0	20	125.6	52	183.2	84
-58.0	-50	14.0	-10	71.6	22	129.2	54	186.8	86
-40.0	-40	17.6	-8	75.2	24	132.8	56	190.4	88
-36.4	-38	21.2	-6	78.8	26	136.4	58	194.0	90
-32.8	-36	24.8	-4	82.4	28	140.0	60	197.6	92
-29.2	-34	28.4	-2	86.0	30	143.6	62	201.2	94
-25.6	-32	32.0	0	89.6	32	147.2	64	204.8	96

1 degree F is 1/180 of the difference between the temperature of melting ice and boiling water.
1 degree C is 1/100 of the difference between the temperature of melting ice and boiling water.

Absolute Zero = -273.16°C = -459.69°F

TABLA DE CONVERSIÓN DE TEMPERATURAS — °F / °C

°F	°C	°F	°C	°F	°C	°F	°C	°F	°C
-459.4	-273	-22.0	-30	35.6	.2	93.2	.34	150.8	.66
-418.0	-250	-18.4	-28	39.2	.4	96	.36	154.4	.68
-328.0	-200	-14.8	-26	42.8	.6	100.4	.38	158.0	.70
-238.0	-150	-11.2	-24	46.4	.8	104.0	.40	161.6	.72
-193.0	-125	-7.6	-22	50.0	.10	107.6	.42	165.2	.74
-148.0	-100	-4.0	-20	53.6	.12	111.2	.44	168.8	.76
-130.0	-90	-0.4	-18	57.2	.14	114.8	.46	172.4	.78
-112.0	-80	3.2	-16	60.8	.16	118.4	.48	176.0	.80
-94.0	-70	6.8	-14	64.4	.18	122.0	.50	179.6	.82
-76.0	-60	10.4	-12	68.0	.20	125.6	.52	183.2	.84
-58.0	-50	14.0	-10	71.6	.22	129.2	.54	186.8	.86
-40.0	-40	17.6	-8	75.2	.24	132.8	.56	190.4	.88
-36.4	-38	21.2	-6	78.8	.26	136.4	.58	194.0	.90
-32.8	-36	24.8	-4	82.4	.28	140.0	.60	197.6	.92
-29.2	-34	28.4	-2	86.0	.30	143.6	.62	201.2	.94
-25.6	-32	32.0	.0	89.6	.32	147.2	.64	204.8	.96

1 grado F es 1/180 de la diferencia entre la temperatura del hielo derretido y la del agua hirviendo
1 grado C es 1/100 de la diferencia entre la temperatura del hielo derretido y la del agua hirviendo

Cero absoluto = - 273.16 °C =-459.69 °F

18-15

CONVERSION TABLE FOR TEMPERATURE – °F/°C (cont.)

°F	°C	°F	°C	°F	°C	°F	°C	°F	°C
208.4	98	347.0	175	590	310	1004	540	6332	3500
212.0	100	356.0	180	608	320	1040	560	7232	4000
221.0	105	365.0	185	626	330	1076	580	4500	8132
230.0	110	374.0	190	644	340	1112	600	9032	5000
239.0	115	383.0	195	662	350	1202	650	9932	5500
248.0	120	392.0	200	680	360	1292	700	10832	6000
257.0	125	410	210	698	370	1382	750	11732	6500
266.0	130	428	220	716	380	1472	800	12632	7000
275.0	135	446	230	734	390	1562	850	13532	7500
284.0	140	464	240	752	400	1652	900	14432	8000
293.0	145	482	250	788	420	1742	950	15332	8500
302.0	150	500	260	824	440	1832	1000	16232	9000
311.0	155	518	270	860	460	2732	1500	17132	9500
320.0	160	536	280	896	480	3632	2000	18032	10000
329.0	165	554	290	932	500	4532	2500		
338.0	170	572	300	968	520	5432	3000		

1 degree F is 1/180 of the difference between the temperature of melting ice and boiling water.
1 degree C is 1/100 of the difference between the temperature of melting ice and boiling water.

Absolute Zero = -273.16°C = -459.69°F

TABLA DE CONVERSIÓN DE TEMPERATURAS — °F / °C (cont.)

°F	°C	°F	°C	°F	°C	°F	°C	°F	°C
208.4 98	347.0 175	590 310	1004 540	6332 3500
212.0 100	356.0 180	608 320	1040 560	7232 4000
221.0 105	365.0 185	626 330	1076 580	4500 8132
230.0 110	374.0 190	644 340	1112 600	9032 5000
239.0 115	383.0 195	662 350	1202 650	9932 5500
248.0 120	392.0 200	680 360	1292 700	10832 6000
257.0 125	410 210	698 370	1382 750	11732 6500
266.0 130	428 220	716 380	1472 800	12632 7000
275.0 135	446 230	734 390	1562 850	13532 7500
284.0 140	464 240	752 400	1652 900	14432 8000
293.0 145	482 250	788 420	1742 950	15332 8500
302.0 150	500 260	824 440	1832 1000	16232 9000
311.0 155	518 270	860 460	2732 1500	17132 9500
320.0 160	536 280	896 480	3632 2000	18032 10000
329.0 165	554 290	932 500	4532 2500	
338.0 170	572 300	968 520	5432 3000	

1 grado F es 1/180 de la diferencia entre la temperatura del hielo derretido y la del agua hirviendo
1 grado C es 1/100 de la diferencia entre la temperatura del hielo derretido y la del agua hirviendo

Cero absoluto = -273.16 °C = -459.69 °F

DECIBEL LEVELS OF SOUNDS

The definition of sound intensity is energy (erg) transmitted per 1 second over a square centimeter surface. Sounds are measured in decibels. A decibel (dB) change of 1 is the smallest change detected by humans.

Hearing Intensity	Decibel Level	Examples of Sounds
Barely Audible	0	Dead silence
		Audible hearing threshold
	10	Room (sound proof)
(Very Light)	20	Empty auditorium
		Ticking of a stopwatch
		Soft whispering
Audible	30	People talking quietly
Light	40	Quiet street noise without autos
Medium	45	Telephone operator
Loud	50	Fax machine in office
	60	Close conversation
Loud	70	Stereo system
		Computer printer
	80	Fire truck/Ambulance siren
		Cat/dog fight
Extremely Loud	90	Industrial machinery
		High school marching band
Damage Possible	100	Heavy duty grinder in a machine/welding shop
Damaging	100+	Begins ear damage
	110	Diesel engine of a train
	120	Lighting strike (thunderstorm)
		60 ton metal forming factory press
	130	60" fan in a bus vacuum system
	140	Commercial/Military jet engine
Ear Drum Shattering	194	Space shuttle engines
	225	16" Guns on a battleship

NIVELES DE SONIDOS EN DECIBELES

La definición de intensidad del sonido es energía (erg) transmitida cada segundo por una superficie de 1 centímetro cuadrado. Los sonidos se miden en decibeles. Un cambio de 1 decibel (db) es el cambio más pequeño detectado por los seres humanos.

Intensidad auditiva	Nivel en decibeles	Ejemplos de sonidos
Apenas audible	0	Silencio absoluto
		Umbral de audición
	10	Habitación (a prueba de sonidos)
(Muy suave)	20	Auditorio vacío
		Tic-tac de un cronómetro
		Murmullo suave
Audible pero suave	30	Gente hablando en voz baja
	40	Ruido de calle tranquila sin autos
Mediana intensidad	45	Operador telefónico
	50	Equipo de fax en una oficina
	60	Conversación cercana
Fuerte	70	Sistema musical estéreo
		Impresora de matriz de puntos de una computadora
	80	Sirena de un camión de bomberos/ambulancia
		Pelea entre perros y/o gatos
Muy fuerte	90	Maquinaria industrial
		Banda musical de escuela secundaria
Con posibilidad de causar daño	100	Afiladora de servicio pesado en un taller de máquinas / soldadura
Dañino	100+	Comienza el daño auditivo
	110	Motor diesel de una locomotora ferroviaria
	120	Caída de un rayo (tormenta eléctrica)
		Prensa de 60 toneladas para dar forma a metales en fábrica
	130	Ventilador de 1.5 m (60") en un sistema de aspiración / vacío
	140	Motor de reacción de un avión comercial o militar
Rotura del tímpano	194	Motores de transbordador espacial
	225	Cañones de 16" en un barco de guerra

SOUND AWARENESS AND SAFETY

Sound Awareness Changes

The typical range of human hearing is 30 hertz – 15,000 hertz. Human hearing recognizes an increase of 20 decibels, such as a stereo sound level increase, as being four times as loud at the higher level than it was at the lower level.

Awareness in Human Hearing	Decibel Change
Noticeably Louder	10
Easily Audible	5
Faintly Audible	3

HEARING PROTECTION LEVELS

Because of the occupational safety and health act of 1970, hearing protection is mandatory if the following time exposures to decibel levels are exceeded because of possible damage to human hearing.

Decibel Level	Time Exposure Per Day
115	15 minutes
110	30 minutes
105	1 hour
102	1-1/2 hours
100	2 hours
97	3 hours
95	4 hours
92	6 hours
90	8 hours

PERCEPCIÓN Y SEGURIDAD SONORAS

Cambios en la percepción de sonidos

El rango típico de la audición humana se encuentra entre 30 hertz y 15,000 hertz. La audición humana reconoce un incremento de 20 decibeles, como el incremento del nivel de sonido de un equipo estéreo, de cuatro veces la intensidad que tenía en el nivel anterior.

Percepción auditiva del ser humano	Cambio en decibeles
Apreciablemente más fuerte	10
Fácilmente perceptible	5
Apenas audible	3

NIVELES DE PROTECCIÓN AUDITIVA

En virtud de la ley de salud y seguridad ocupacional de 1970, es obligatoria la protección auditiva si se excede el siguiente tiempo de exposición a determinados niveles de decibeles debido a posibles daños a la audición humana.

Nivel en decibeles	Tiempo diario de exposición
115	15 minutos
110	30 minutos
105	1 hora
102	1-1/2 horas
100	2 horas
97	3 horas
95	4 horas
92	6 horas
90	8 horas

TRIGONOMETRIC FORMULAS — RIGHT TRIANGLE

Angles = X, Y, Z
Distances = x, y, z

Area $= \dfrac{x\,y}{2}$

$\sin X = \dfrac{x}{z}$ $\cos X = \dfrac{y}{z}$

$\tan X = \dfrac{x}{y}$ $\cot X = \dfrac{y}{x}$

Pythagorean Theorem states

That $x^2 + y^2 = z^2$

Thus $x = \sqrt{z^2 - y^2}$

Thus $y = \sqrt{z^2 - x^2}$

Thus $z = \sqrt{x^2 + y^2}$

Given X and z, find Y, x and y

$Y = 90^\circ - X$, $x = z \sin X$, $y = z \cos X$

Given X and z, find Y, x and y

$\sin X = \dfrac{x}{z} = \cos Y$, $y = \sqrt{(z^2 - x^2)} = z\sqrt{1 - \dfrac{x^2}{z^2}}$

Given X and z, find Y, x and z

$Y = 90^\circ - X$, $x = y \tan X$, $z = \dfrac{y}{\cos X}$

Given x and y, find X, Y and z

$\tan X = \dfrac{x}{y} = \cot Y$, $z = \sqrt{x^2 + y^2} = x\sqrt{1 + \dfrac{y^2}{x^2}}$

Given X and x, find Y, y and z

$Y = 90^\circ - X$, $y = x \cot X$, $z = \dfrac{x}{\sin X}$

FÓRMULAS TRIGONOMÉTRICAS – TRIÁNGULO RECTÁNGULO

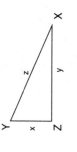

Ángulos = X, Y, Z
Distancias = x, y, z
Superficie = $\dfrac{x \cdot y}{2}$

$$\operatorname{sen} X = \frac{x}{z} \qquad \cos X = \frac{y}{z}$$

$$\tan X = \frac{x}{y} \qquad \cot X = \frac{y}{x}$$

El teorema de Pitágoras establece que $\quad x^2 + y^2 = z^2$

Por lo tanto $x = \sqrt{z^2 - y^2}$

Por lo tanto $y = \sqrt{z^2 - x^2}$

Por lo tanto $z = \sqrt{x^2 + y^2}$

Dados x y z, encontrar X, Y e y

$$\operatorname{Sen} X = \frac{x}{z} = \cos Y, \quad y = \sqrt{(z^2 - x^2)} = z\sqrt{1 - \frac{x^2}{z^2}}$$

Dados x e y, encontrar X, Y y z

$$\operatorname{Tan} X = \frac{x}{y} = \cot Y, \quad z = \sqrt{x^2 + y^2} = x\sqrt{1 + \frac{y^2}{x^2}}$$

Dados X y x, encontrar Y, y y z

$$Y = 90^{o} - X, \quad y = x \cot X, \quad z = \frac{x}{\operatorname{sen} X}$$

Dados X y z, encontrar Y, x e y

$$Y = 90^{o} - X, \quad x = z \operatorname{sen} X, \quad y = z \cos X$$

Dados X y z, encontrar Y, x y z

$$Y = 90^{o} - X, \quad x = y \tan X, \quad z = \frac{y}{\cos X}$$

TRIGONOMETRIC FORMULAS — OBLIQUE TRIANGLES

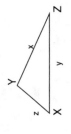

Given x, y, and z, find X, Y and z

$$X + Y = 180^\circ - Z, \quad z = \frac{x \sin Z}{\sin X}, \quad \tan X = \frac{x \sin Z}{y - (x \cos Z)}$$

Given X, x and y, Find Y, Z and z

$$\sin Y = \frac{y \sin X}{x}, \quad Z = 180^\circ - (X+Y), \quad z = \frac{x \sin Z}{\sin X}$$

Given X, Y and x, Find y, Z and z

$$y = \frac{x \sin Y}{\sin X}, \quad Z = 180^\circ - (X+Y), \quad z = \frac{x \sin Z}{\sin X}$$

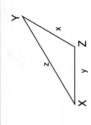

Given x, y and z, Find X, Y and Z

$$s = \frac{x + y + z}{2}, \quad \sin \tfrac{1}{2} X = \sqrt{\frac{(s-y)(s-z)}{yz}}$$

$$\sin \tfrac{1}{2} Y = \sqrt{\frac{(s-x)(s-z)}{xz}}, \quad C = 180^\circ - (X+Y)$$

Given x, y and z, find the Area

$$s = \frac{x + y + z}{2}, \quad \text{Area} = \sqrt{S(s-x)(s-y)(s-z)}$$

$$\text{Area} = \frac{yz \sin X}{2}, \quad \text{Area} = \frac{x^2 \sin Y \sin Z}{2 \sin X}$$

18-24

FÓRMULAS TRIGONOMÉTRICAS - TRIÁNGULOS OBLICUOS

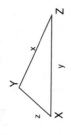

Dados x, y y z, encontrar X, Y y Z

$$s = \frac{x+y+z}{2} \ , \ \operatorname{sen}\frac{1}{2} \ X = \sqrt{\frac{(s-y)(s-z)}{yz}}$$

$$\operatorname{sen}\frac{1}{2} \ Y = \sqrt{\frac{(s-x)(s-z)}{xz}} \ , \ C = 180^{\circ} - (X+Y)$$

Dados x, y y z, encontrar la superficie

$$s = \frac{x+y+z}{2} \ , \ \text{Superficie} = \sqrt{S(s-x)(s-y)(s-z)}$$

$$\text{Superficie} = \frac{yz \operatorname{sen} X}{2} \ , \ \text{Superficie} = \frac{x^2 \operatorname{sen} Y \operatorname{sen} Z}{2 \operatorname{sen} X}$$

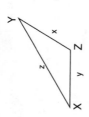

Dados x, y, y Z, encontrar X, Y y z

$$X+Y = 180^{\circ} - Z, \ z = \frac{x \operatorname{sen} Z}{\operatorname{sen} X} \ , \ \tan X = \frac{x \operatorname{sen} Z}{y - (x \cos Z)}$$

Dados X, x, e y, encontrar Y, Z e z

$$\operatorname{sen} Y = \frac{y \operatorname{sen} X}{x} \ , \ Z = 180^{\circ} - (X+Y), \ z = \frac{x \operatorname{sen} Z}{\operatorname{sen} X}$$

Dados X, Y, y x, encontrar Y, Z y z

$$y = \frac{x \operatorname{sen} Y}{\operatorname{sen} X} \ , \ Z = 180^{\circ} - (X+Y) \ , \ z = \frac{x \operatorname{sen} Z}{\operatorname{sen} X}$$

TRIGONOMETRIC FORMULAS — SHAPES

Equilateral Triangle	Annulus	Trapezium

Equilateral Triangle

X = Sides (Equal Lengths)

$$Area = X^2 \frac{\sqrt{3}}{4} = .433\, X^2$$

$$Perimeter = 3\, X$$

$$H = \frac{X}{2} \sqrt{3} = .866\, X$$

Annulus

C_1 and R_1 = Inside Circle

C_2 and R_2 = Outside Circle

C = Circumference

R = Radius

$$Area = \pi(R_1 + R_2)(R_2 - R_1)$$

$$Area = \left((C_2)^2 - (C_1)^2\right).7854$$

Trapezium

Perimeter is the

Sum of L, M, N and O

$$Area = \frac{(S+T)\,Q + RS + PT}{2}$$

FÓRMULAS TRIGONOMÉTRICAS – FORMAS

Triángulo equilátero	Corona circular	Trapezoide

Triángulo equilátero

X = lados (Iguales longitudes)

$Superficie = X^2 \sqrt{\dfrac{3}{4}} = .433\, X^2$

$Perímetro = 3\, X$

$H = \dfrac{X}{2} \sqrt{3} = .866\, X$

Corona circular

C_1 y R_1 = Círculo interior

C_2 y R_2 = Círculo exterior

C = Circunferencia

R = Radio

$Superficie = \pi (R_1 + R_2)(R_2 - R_1)$

$Superficie = \left((C_2)^2 - (C_1)^2 \right) .7854$

Trapezoide

El perímetro es la suma de L, M, N y O

$Superficie = \dfrac{(S + T)\, Q + RS + PT}{2}$

18-27

TRIGONOMETRIC FORMULAS — SHAPES (cont.)

Trapezoid

Perimeter =
The Sum of the lengths of all four sides

$$Area = \frac{(X + Y)}{2}$$

Quadrilateral

$$Area = \frac{L_1 \cdot L_2 \cdot Sin\ \theta}{2}$$

Where θ = Degrees of Angle

Rectangle

Area = XY
Diagonal Line (D)

$$= \sqrt{X^2 + Y^2}$$

Perimeter =
2(X + Y)
If a square then X = Y

Parallelogram

Where θ = Degrees of Angle

Area =
XH = XY sin θ
Perimeter =
2(X + Y)

FÓRMULAS TRIGONOMÉTRICAS – FORMAS (cont.)

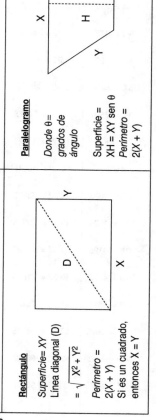

Trapezoide

Perímetro = La suma de las longitudes de los cuatro lados

$$Superficie = \frac{(X + Y)}{2}$$

Rectángulo

Superficie = XY
Línea diagonal (D)
$$= \sqrt{X^2 + Y^2}$$
Perímetro =
$2(X + Y)$
Si es un cuadrado, entonces $X = Y$

Cuadrilátero

$$Superficie = \frac{L_1 \cdot L_2 \cdot Sen\ \theta}{2}$$

Donde θ = grados de ángulo

Paralelogramo

Donde θ = grados de ángulo

Superficie =
$XH = XY\ sen\ \theta$
Perímetro =
$2(X + Y)$

18-29

DECIMAL EQUIVALENTS OF FRACTIONS

8ths	32nds	64ths	64ths
1/8 = .125	1/32 = .03125	1/64 = 0.15625	33/64 = .515625
1/4 = .250	3/32 = .09375	3/64 = .046875	35/64 = .546875
3/8 = .375	5/32 = .15625	5/64 = .078125	37/64 = .57812
1/2 = .500	7/32 = .21875	7/64 = .109375	39/64 = .609375
5/8 = .625	9/32 = .28125	9/64 = .140625	41/64 = .640625
3/4 = .750	11/32 = .34375	11/64 = .171875	43/64 = .671875
7/8 = .875	13/32 = .40625	13/64 = .203128	45/64 = .703125
16ths	15/32 = .46875	15/64 = .234375	47/64 = .734375
1/16 = .0625	17/32 = .53125	17/64 = .265625	49/64 = .765625
3/16 = .1875	19/32 = .59375	19/64 = .296875	51/64 = 3796875
5/16 = .3125	21/32 = .65625	21/64 = .328125	53/64 = .828125
7/16 = .4375	23/32 = .71875	23/64 = .359375	55/64 = .859375
9/16 = .5625	25/32 = .78125	25/64 = .390625	57/64 = .890625
11/16 = .6875	27/32 = .84375	27/64 = .421875	59/64 = .921875
13/16 = .8125	29/32 = .90625	29/64 = .453125	61/64 = .953125
15/16 = .9375	31/32 = .96875	31/64 = .484375	63/64 = .984375

EQUIVALENTES DECIMALES DE FRACCIONES

8avos	32avos	64avos	64avos
1/8 = .125	1/32 = .03125	1/64 = 0.15625	33/64 = .515625
1/4 = .250	3/32 = .09375	3/64 = .046875	35/64 = .546875
3/8 = .375	5/32 = .15625	5/64 = .078125	37/64 = .57812
1/2 = .500	7/32 = .21875	7/64 = .109375	39/64 = .609375
5/8 = .625	9/32 = .28125	9/64 = .140625	41/64 = .640625
3/4 = .750	11/32 = .34375	11/64 = .171875	43/64 = .671875
7/8 = .875	13/32 = .40625	13/64 = .203128	45/64 = .703125
16avos	15/32 = .46875	15/64 = .234375	47/64 = .734375
1/16 = .0625	17/32 = .53125	17/64 = .265625	49/64 = .765625
3/16 = .1875	19/32 = .59375	19/64 = .296875	51/64 = 3796875
5/16 = .3125	21/32 = .65625	21/64 = .328125	53/64 = .828125
7/16 = .4375	23/32 = .71875	23/64 = .359375	55/64 = .859375
9/16 = .5625	25/32 = .78125	25/64 = .390625	57/64 = .890625
11/16 = .6875	27/32 = .84375	27/64 = .421875	59/64 = .921875
13/16 = .8125	29/32 = .90625	29/64 = .453125	61/64 = .953125
15/16 = .9375	31/32 = .96875	31/64 = .484375	63/64 = .984375

COMMON ENGINEERING UNITS AND THEIR RELATIONSHIP

Quantity	SI Metric Units/Symbols	Customary Units	Relationship of Units
Acceleration	meters per second squared (m/s²)	feet per second squared (ft/s²)	m/s² = ft/s² x 3.281
Area	square meter (m²) square millimeter (mm²)	square foot (ft²) square inch (in²)	m² = ft² x 10.764 mm² = in² x 0.00155
Density	kilograms per cubic meter (kg/m³) grams per cubic centimeter (g/cm³)	pounds per cubic foot (lb/ft³) pounds per cubic inch (lb/in³)	kg/m³ = lb/ft³ x 16.02 g/cm³ = lb/in³ x 0.036
Work	Joule (J)	foot pound force (ft lbf or ft lb)	J = ft lbf x 1.356
Heat	Joule (J)	British thermal unit (Btu) Calorie (Cal)	J = Btu x 1.055 J = cal x 4.187
Energy	kilowatt (kW)	Horsepower (HP)	kW = HP x 0.7457
Force	Newton (N) Newton (N)	Pound-force (lbf, lb · f, or lb) kilogram-force (kgf, kg · f, or kp)	N = lbf x 4.448 N = $\frac{kgf}{9.807}$
Length	meter (m) millimeter (mm)	foot (ft) inch (in)	m = ft x 3.281 mm = $\frac{in}{25.4}$
Mass	kilogram (kg) gram (g)	pound (lb) ounce (oz)	kg = lb x 2.2 g = $\frac{oz}{28.35}$
Stress	Pascal = Newton per second (Pa = N/s)	pounds per square inch (lb/in² or psi)	Pa = lb/in² x 6,895
Temperature	degree Celsius (°C)	degree Fahrenheit (°F)	°C = $\frac{°F - 32}{1.8}$
Torque	Newton meter (N · m)	foot-pound (ft lb) inch-pound (in lb)	N · m = ft lbf x 1.356 N · m = in lbf x 0.113
Volume	cubic meter (m³) cubic centimeter (cm³)	cubic foot (ft³) cubic inch (in³)	m³ = ft³ x 35.314 cm³ = $\frac{in^3}{16.387}$

UNIDADES COMUNES DE INGENIERÍA Y SUS RELACIONES

Cantidad	Unidades/Símbolos métricos SI	Unidades convencionales	Relación de unidades
Aceleración	metros por segundo al cuadrado (m/s^2)	pies por segundo al cuadrado (ft/s^2)	m/s^2 = ft/s^2 x 3.281
Superficie	metro cuadrado (m^2) milímetro cuadrado (mm^2)	pie cuadrado (ft^2) pulgada cuadrada (in^2)	m^2 = ft^2 x 10.764 mm^2 = in^2 x 0.00155
Densidad	kilogramos por metro cúbico (kg/m^3) gramos por centímetro cúbico (g/cm^3)	libras por pie cúbico (lb/ft^3) libras por pulgada cúbica (lb/in^3)	kg/m^3 = lb/ft^3 x 16.02 g/cm^3 = lb/in^3 x 0.036
Trabajo	Joule (J)	pie libras fuerza (ft lbf o ft lb)	J = ft lbf x 1.356
Calor	Joule (J)	Unidad térmica del sistema inglés (Btu) Caloría (Cal)	J = Btu x 1.055 J = cal x 4.187
Energía	kilowatt (kW)	Potencia (HP)	kW = HP x 0.7457
Fuerza	Newton (N) Newton (N)	libra-fuerza (lbf, lb · f, or lb) kilogramo-fuerza (kgf, kg · f, or kp)	N = lbf x 4.448 N = $\dfrac{kgf}{9.807}$
Longitud	metro (m) milímetro (mm)	pie (ft) pulgada (in)	m = ft x 3.281 mm = $\dfrac{in}{25.4}$
Masa	kilogramo (kg) gramo (g)	libra (lb) onza (oz)	kg = lb x 2.2 g = $\dfrac{oz}{28.35}$
Voltaje	Pascal = Newton por segundo (Pa = N/s)	libras por pulgada cuadrada (lb/in^2 o psi)	Pa = lb/in^2 x 6,895
Temperatura	grado Celsius (°C)	grado Fahrenheit (°F)	°C = $\dfrac{°F - 32}{1.8}$
Par motor	Newton metro (N · m)	libra-pie (ft lb) pulgada-libra (in lb)	N · m = ft lbf x 1.356 N · m = in lbf x 0.113
Volumen	metro cúbico (m^3) centímetro cúbico (cm^3)	pie cúbico (ft^3) pulgada cúbica (in^3)	m^3 = ft^3 x 35.314 cm^3 = $\dfrac{in^3}{16.387}$

COMMONLY USED GEOMETRICAL RELATIONSHIPS

Diameter of a circle × 3.1416 = Circumference

Radius of a circle × 6.283185 = Circumference

Square of the radius of a circle × 3.1416 = Area

Square of the diameter of a circle × 0.7854 = Area

Square of the circumference of a circle × 0.07958 = Area

Half the circumference of a circle × half its diameter = Area

Circumference of a circle × 0.159155 = Radius

Square root of the area of a circle × 0.56419 = Radius

Circumference of a circle × 0.31831 = Diameter

Square root of the area of a circle × 1.12838 = Diameter

Diameter of a circle × 0.866 = Side of an inscribed equilateral triangle

Diameter of a circle × 0.7071 = Side of an inscribed square

Circumference of a circle × 0.225 = Side of an inscribed square

Circumference of a circle × 0.282 = Side of an equal square

Diameter of a circle × 0.8862 = Side of an equal square

Base of a triangle × one-half the altitude = Area

Multiplying both diameters and .7854 together = Area of an ellipse

Surface of a sphere × one-sixth of its diameter = Volume

Circumference of a sphere × its diameter = Surface

Square of the diameter of a sphere × 3.1416 = Surface

Square of the circumference of a sphere × 0.3183 = Surface

Cube of the diameter of a sphere × 0.5236 = Volume

Cube of the circumference of a sphere × 0.016887 = Volume

Radius of a sphere × 1.1547 = Side of an inscribed cube

Diameter of a sphere divided by $\sqrt{3}$ = Side of an inscribed cube

Area of its base × one-third of its altitude = Volume of a cone or
pyramid whether round, square or triangular

Area of one of its sides × 6 = Surface of the cube

Altitude of trapezoid × one-half the sum of its parallel sides = Area

RELACIONES GEOMÉTRICAS DE USO COMÚN

Diámetro de un círculo x 3.1416 = Circunferencia.

Radio de un círculo x 6.283185 = Circunferencia.

Cuadrado del radio de un círculo x 3.1416 = Superficie.

Cuadrado del diámetro de un círculo x 0.7854 = Superficie.

Cuadrado de la circunferencia de un círculo x 0.07958 = Superficie.

Semicircunferencia de un círculo x la mitad de su diámetro = Superficie.

Circunferencia de un círculo x 0.159155 = Radio.

Raíz cuadrada de la superficie de un círculo x 0.56419 = Radio.

Circunferencia de un círculo x 0.31831 = Diámetro.

Raíz cuadrada de la superficie de un círculo x 1.12838 = Diámetro.

Diámetro de un círculo x 0.866 = Lado de un triángulo equilátero inscrito.

Diámetro de un círculo x 0.7071 = Lado de un cuadrado inscrito.

Circunferencia de un círculo x 0.225 = Lado de un cuadrado inscrito.

Circunferencia de un círculo x 0.282 = Lado de un cuadrado equivalente.

Diámetro de un círculo x 0.8862 = Lado de un cuadrado igual.

Base de un triángulo x la mitad de su altura = Superficie.

Multiplicación de ambos diámetros entre sí y luego por .7854 = Superficie de una elipse.

Superficie de una esfera x un sexto de su diámetro = Volumen.

Circunferencia de una esfera x su diámetro = Superficie.

Cuadrado del diámetro de una esfera x 3.1416 = Superficie.

Cuadrado de la circunferencia de una esfera x 0.3183 = Superficie.

Cubo del diámetro de una esfera x 0.5236 = Volumen.

Cubo de la circunferencia de una esfera x 0.016887 = Volumen.

Radio de una esfera x 1.1547 = Lado de un cubo inscrito.

Diámetro de una esfera dividido por $\sqrt{3}$ = Lado de un cubo inscrito.

Superficie de su base x un tercio de su altura = Volumen de un cono o pirámide ya sean redondos, cuadrados o triangulares.

Superficie de uno de sus lados x 6 = Superficie del cubo.

Altura del trapezoide x la semisuma de sus lados paralelos = Superficie.

NOTES — NOTAS

CHAPTER 19/CAPITULO 19
Materials and Tools
Materiales y Herramientas

In this chapter you will find technical information on materials and tools. These are shown with all-English on the left-hand page and all-Spanish on the right page. The two pages will almost always be identical. So, the fifth line from the bottom on the left will convey exactly the same information as the fifth line from the bottom on the right.

En éste capítulo usted encontrará la información técnica sobre los materiales y las herramientas. Éstos se demuestran con todo-Inglés en la página izquierda y todo-Español en la página derecha. Las dos páginas serán siempre idénticas. Así pues, la quinta línea del fondo a la izquierda transportará exactamente la misma información que la quinta línea del fondo a la derecha.

SHEET METAL SCREW CHARACTERISTICS

Screw Size #	Screw Dia. (Inches)	Diameter of Pierced Hole (Inches)	Hole Size #	Thickness of Metal – Gauge #
4	.112	.086	44	28
		.086	44	26
		.093	42	24
		.098	42	22
		.100	40	20
6	.138	.111	39	28
		.111	39	26
		.111	39	24
		.111	38	22
		.111	36	20
7	.155	.121	37	28
		.121	37	26
		.121	35	24
		.121	33	22
		.121	32	20
		–	31	18
8	.165	.137	33	26
		.137	33	24
		.137	32	22
		.137	31	20
		–	30	18
10	.191	.158	30	26
		.158	30	24
		.158	30	22
		.158	29	20
		.158	25	18
12	.218	–	26	24
		.185	25	22
		.185	24	20
		.185	22	18
14	.251	–	15	24
		.212	12	22
		.212	11	20
		.212	9	18

Deviations in materials and conditions could require variations from these dimensions.

CARACTERÍSTICAS DE TORNILLO DE PLANCHA

Num. de tamaño del tornillo	Diám. de tornillo (pulgadas)	Diámetro de agujero perforado (pulgadas)	Tamaño del agujero	Núm. de espesor del metal – calibre
#4	.112	.086	#44	28
		.086	#44	26
		.093	#42	24
		.098	#42	22
		.100	#40	20
#6	.138	.111	#39	28
		.111	#39	26
		.111	#39	24
		.111	#38	22
		.111	#36	20
#7	.155	.121	#37	28
		.121	#37	26
		.121	#35	24
		.121	#33	22
		.121	#32	20
		–	#31	18
#8	.165	.137	#33	26
		.137	#33	24
		.137	#32	22
		.137	#31	20
		–	#30	18
#10	.191	.158	#30	26
		.158	#30	24
		.158	#30	22
		.158	#29	20
		.158	#25	18
#12	.218	–	#26	24
		.185	#25	22
		.185	#24	20
		.185	#22	18
#14	.251	–	#15	24
		.212	#12	22
		.212	#11	20
		.212	#9	18

Las desviaciones en los materiales y las condiciones podrían requerir variaciones en estas dimensiones.

STANDARD WOOD SCREW CHARACTERISTICS (INCHES)

Screw Size #	Wood Screw Standard Lengths (Inches)	Size of Pilot Hole Softwood Bit #	Size of Pilot Hole Hardwood Bit #	Size of Shank Hole Clearance Bit #	Size of Shank Hole Hole Diameter (Inches)
0	¼	75	66	52	.060
1	¼ to ⅜	71	57	47	.073
2	¼ to ½	65	54	42	.086
3	¼ to ⅝	58	53	37	.099
4	⅜ to ¾	55	51	32	.112
5	⅜ to ¾	53	47	30	.125
6	⅜ to 1½	52	44	27	.138
7	⅜ to 1½	51	39	22	.151
8	½ to 2	48	35	18	.164
9	⅝ to 2¼	45	33	14	.177
10	⅝ to 2½	43	31	10	.190
11	¾ to 3	40	29	4	.203
12	⅞ to 3½	38	25	2	.216
14	1 to 4½	32	14	D	.242
16	1¼ to 5½	29	10	I	.268
18	1½ to 6	26	6	N	.294
20	1¾ to 6	19	3	P	.320
24	3½ to 6	15	D	V	.372

CARACTERÍSTICAS DE LOS TORNILLOS NORMALES PARA MADERA (PULGADAS)

Núm. de tamaño del tornillo	Longitudes normales de tornillos para madera	Tamaño del agujero piloto		Tamaño del agujero del cuerpo	
		Núm. de broca para madera blanda	Núm. de broca para madera dura	Núm. de broca de resguardo	Diám. del agujero
0	¼	75	66	52	.060
1	¼ to ⅜	71	57	47	.073
2	¼ to ½	65	54	42	.086
3	¼ to ⅝	58	53	37	.099
4	⅜ to ¾	55	51	32	.112
5	⅜ to ¾	53	47	30	.125
6	⅜ to 1½	52	44	27	.138
7	⅜ to 1½	51	39	22	.151
8	½ to 2	48	35	18	.164
9	⅝ to 2¼	45	33	14	.177
10	⅝ to 2½	43	31	10	.190
11	¾ to 3	40	29	4	.203
12	⅞ to 3½	38	25	2	.216
14	1 to 4½	32	14	D	.242
16	1¼ to 5½	29	10	I	.268
18	1½ to 6	26	6	N	.294
20	1¾ to 6	19	3	P	.320
24	3½ to 6	15	D	V	.372

ALLEN HEAD AND MACHINE SCREW BOLT AND TORQUE CHARACTERISTICS

Number of Threads Per Inch	Allen Head And Mach. Screw Bolt Size	Allen Head Case H Steel 160,000 psi	Mach. Screw Yellow Brass 60,000 psi	Mach. Screw Silicone Bronze 70,000 psi
		Torque in Foot-Pounds or Inch-Pounds		
4.5	2"	8800	–	–
5	1¾"	6100	–	–
6	1½"	3450	655	595
6	1⅜"	2850	–	–
7	1¼"	2130	450	400
7	1⅛"	1520	365	325
8	1"	970	250	215
9	⅞"	640	180	160
10	¾"	400	117	104
11	⅝"	250	88	78
12	9⁄16"	180	53	49
13	½"	125	41	37
14	7⁄16"	84	30	27
16	⅜"	54	20	17
18	5⁄16"	33	125 in#	110 in#
20	¼"	16	70 in#	65 in#
24	#10	60	22 in#	20 in#
32	#8	46	19 in#	16 in#
32	#6	21	10 in#	8 in#
40	#5	–	7.2 in#	6.4 in#
40	#4	–	4.9 in#	4.4 in#
48	#3	–	3.7 in#	3.3 in#
56	#2	–	2.3 in#	2 in#

For fine thread bolts, increase by 9%.

CARACTERÍSTICAS DE PERNOS Y TORSIÓN PARA TORNILLOS CON CABEZA ALLEN Y PARA METALES

Número de roscas por pulgada	Tamaño de perno cabeza Allen y para metales	Cabeza Allen de acero cementado 160,000 PSI	Tornillo para metales bronce amarillo 60,000 psi	Tornillo para metales bronce siliconado 70,000 psi
		Torsión en pies-libra o pulgadas-libra		
4.5	2"	8800	–	–
5	1¾"	6100	–	–
6	1½"	3450	655	595
6	1⅜"	2850	–	–
7	1¼"	2130	450	400
7	1⅛"	1520	365	325
8	1"	970	250	215
9	⅞"	640	180	160
10	¾"	400	117	104
11	⅝"	250	88	78
12	⁹⁄₁₆"	180	53	49
13	½"	125	41	37
14	⁷⁄₁₆"	84	30	27
16	⅜"	54	20	17
18	⁵⁄₁₆"	33	125 in#	110 in#
20	¼"	16	70 in#	65 in#
24	#10	60	22 in#	20 in#
32	#8	46	19 in#	16 in#
32	#6	21	10 in#	8 in#
40	#5	–	7.2 in#	6.4 in#
40	#4	–	4.9 in#	4.4 in#
48	#3	–	3.7 in#	3.3 in#
56	#2	–	2.3 in#	2 in#

Para pernos de rosca fina, aumente un 9%.

HEX HEAD BOLT AND TORQUE CHARACTERISTICS

BOLT MAKE-UP IS STEEL WITH COARSE THREADS

Number of Threads Per Inch	Hex Head Bolt Size (Inches)	 SAE 0-1-2 74,000 psi	 SAE Grade 3 100,000 psi	 SAE Grade 5 120,000 psi
		Torque = Foot-Pounds		
4.5	2	2750	5427	4550
5	1¾	1900	3436	3150
6	1½	1100	1943	1775
6	1⅜	900	1624	1500
7	1¼	675	1211	1105
7	1⅛	480	872	794
8	1	310	551	587
9	⅞	206	372	382
10	¾	155	234	257
11	⅝	96	145	154
12	9⁄16	69	103	114
13	½	47	69	78
14	7⁄16	32	47	54
16	⅜	20	30	33
18	5⁄16	12	17	19
20	¼	6	9	10

For fine thread bolts, increase by 9%.

CARACTERÍSTICAS DE PERNOS DE CABEZA HEXAGONAL Y TORSIONES

EL PERNO ESTÁ CONSTRUIDO DE ACERO CON ROSCA GRUESA

Número de roscas por pulgada	Tamaño de perno de cabeza hexagonal	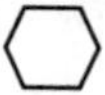 SAE 0-1-2 74,000 psi	 SAE grado 3 100,000 psi	 SAE grado 5 120,000 psi
		Torsión = Libras pie		
4.5	2"	2750	5427	4550
5	1¾"	1900	3436	3150
6	1½"	1100	1943	1775
6	1⅜"	900	1624	1500
7	1¼"	675	1211	1105
7	1⅛"	480	872	794
8	1"	310	551	587
9	⅞"	206	372	382
10	¾"	155	234	257
11	⅝"	96	145	154
12	9⁄16"	69	103	114
13	½"	47	69	78
14	7⁄16"	32	47	54
16	⅜"	20	30	33
18	5⁄16"	12	17	19
20	¼"	6	9	10

Para pernos de rosca fina, aumente un 9%.

HEX HEAD BOLT AND TORQUE CHARACTERISTICS (cont.)

BOLT MAKE-UP IS STEEL WITH COARSE THREADS

Number of Threads Per Inch	Hex Head Bolt Size (Inches)	SAE Grade 6 133,000 psi	SAE Grade 7 133,000 psi	SAE Grade 8 150,000 psi
		Torque = Foot-Pounds		
4.5	2	7491	7500	8200
5	1¾	5189	5300	5650
6	1½	2913	3000	3200
6	1⅜	2434	2500	2650
7	1¼	1815	1825	1975
7	1⅛	1304	1325	1430
8	1	825	840	700
9	⅞	550	570	600
10	¾	350	360	380
11	⅝	209	215	230
12	9⁄16	150	154	169
13	½	106	110	119
14	7⁄16	69	71	78
16	⅜	43	44	47
18	5⁄16	24	25	29
20	¼	12.5	13	14

For fine thread bolts, increase by 9%.
For special alloy bolts, obtain torque rating from the manufacturer.

CARACTERÍSTICAS DE PERNOS DE CABEZA HEXAGONAL Y TORSIONES (cont.)

EL PERNO ESTÁ CONSTRUIDO DE ACERO CON ROSCA GRANDE

Número de roscas por pulgada	Tamaño de perno de cabeza hexagonal	SAE grado 6 133,000 psi	SAE grado 7 133,000 psi	SAE grado 8 150,000 psi
		Torsión = Libras pie		
4.5	2"	7491	7500	8200
5	1¾"	5189	5300	5650
6	1½"	2913	3000	3200
6	1⅜"	2434	2500	2650
7	1¼"	1815	1825	1975
7	1⅛"	1304	1325	1430
8	1"	825	840	700
9	⅞"	550	570	600
10	¾"	350	360	380
11	⅝"	209	215	230
12	⁹⁄₁₆"	150	154	169
13	½"	106	110	119
14	⁷⁄₁₆"	69	71	78
16	⅜"	43	44	47
18	⁵⁄₁₆"	24	25	29
20	¼"	12.5	13	14

Para pernos de rosca fina, aumente un 9%.
Para pernos especiales de aleación, obtener la especificación de torsión del fabricante.

WHITWORTH HEX HEAD BOLT AND TORQUE CHARACTERISTICS

BOLT MAKE-UP IS STEEL WITH COARSE THREADS

Number of Threads Per Inch	Whitworth Type Hex Head Bolt Size (Inches)	Grades A & B 62,720 psi	Grade S 112,000 psi	Grade T 123,200 psi	Grade V 145,600 psi
			Torque = Foot-Pounds		
8	1	276	497	611	693
9	7/8	186	322	407	459
11	3/4	118	213	259	287
11	5/8	73	128	155	175
12	9/16	52	94	111	128
12	1/2	36	64	79	89
14	7/16	24	43	51	58
16	3/8	15	27	31	36
18	5/16	9	15	18	21
20	1/4	5	7	9	10

For fine thread bolts, increase by 9%.

CARACTERÍSTICAS DE PERNOS DE CABEZA WHITWORTH HEXAGONAL Y TORSIONES

EL PERNO ESTÁ CONSTRUIDO DE ACERO CON ROSCA GRUESA

Número de roscas por pulgada	Tamaño de perno de cabeza hexagonal tipo Whitworth	Grados A & B 62,720 psi	Grado S 112,000 psi	Grado T 123,200 psi	Grado V 145,600 psi
			Torsión = Libras pie		
8	1"	276	497	611	693
9	7/8"	186	322	407	459
11	3/4"	118	213	259	287
11	5/8"	73	128	155	175
12	9/16"	52	94	111	128
12	1/2"	36	64	79	89
14	7/16"	24	43	51	58
16	3/8"	15	27	31	36
18	5/16"	9	15	18	21
20	1/4"	5	7	9	10

Para pernos de rosca fina, aumente un 9%.

19-13

METRIC HEX HEAD BOLT AND TORQUE CHARACTERISTICS

BOLT MAKE-UP IS STEEL WITH COARSE THREADS

Metric Type Thread Pitch	(Dimensions in Millimeters) Bolt Size	5D Standard 5D 71,160 psi	8G Standard 8G 113,800 psi	10K Standard 10K 142,000 psi	12K Standard 12K 170,674 psi
		Torque = Foot-Pounds			
3.0	24	261	419	570	689
2.5	22	182	284	394	464
2.0	18	111	182	236	183
2.0	16	83	132	175	208
1.25	14	55	89	117	137
1.25	12	34	54	70	86
1.25	10	19	31	40	49
1.0	8	10	16	22	27
1.0	6	5	6	8	10

For fine thread bolts, increase by 9%.

19-14

CARACTERÍSTICAS DE PERNOS DE CABEZA HEXAGONAL Y TORSIONES EN UNIDADES MÉTRICAS

EL PERNO ESTÁ CONSTRUIDO DE ACERO CON ROSCA GRUESA

(Tipo métrico) Paso de la rosca	(Dimensiones en milímetros) Tamaño de perno	5D Estándar 5D 71,160 psi	8G Estándar 8G 113,800 psi	10K Estándar 10K 142,000 psi	12K Estándar 12K 170,674 psi
			Torsión = Libras pie		
3.0	24	261	419	570	689
2.5	22	182	284	394	464
2.0	18	111	182	236	183
2.0	16	83	132	175	208
1.25	14	55	89	117	137
1.25	12	34	54	70	86
1.25	10	19	31	40	49
1.0	8	10	16	22	27
1.0	6	5	6	8	10

Para pernos de rosca fina, aumente un 9%.

19-15

TIGHTENING TORQUE IN POUND-FEET-SCREW FIT

Wire Size, AWG/kcmil	Driver	Bolt	Other
18-16	1.67	6.25	4.2
14-8	1.67	6.25	6.125
6-4	3.0	12.5	8.0
3-1	3.2	21.00	10.40
0-2/0	4.22	29	12.5
3/0-200	–	37.5	17.0
250-300	–	50.0	21.0
400	–	62.5	21.0
500	–	62.5	25.0
600-750	–	75.0	25.0
800-1000	–	83.25	33.0
1250-2000	–	83.26	42.0

SCREW TORQUES

Screw Size, Inches Across, Hex Flats	Torque, Pound-Feet
1/8	4.2
5/32	8.3
3/16	15
7/32	23.25
1/4	42

TORSIÓN DE AJUSTE EN LIBRAS-PIE, SEGÚN TORNILLO

Tamaño de cable, AWG/kcmil	Impulsor	Perno	Otros
18-16	1.67	6.25	4.2
14-8	1.67	6.25	6.125
6-4	3.0	12.5	8.0
3-1	3.2	21.00	10.40
0-2/0	4.22	29	12.5
3/0-200	–	37.5	17.0
250-300	–	50.0	21.0
400	–	62.5	21.0
500	–	62.5	25.0
600-750	–	75.0	25.0
800-1000	–	83.25	33.0
1250-2000	–	83.26	42.0

TORSIONES DE TORNILLOS

Tamaño del tornillo, pulgadas lado a lado, planas hexagonales	Torsión, libras-pie
1/8	4.2
5/32	8.3
3/16	15
7/32	23.25
1/4	42

STANDARD TAPS AND DIES (IN INCHES)

Thread Size	Coarse			Fine		
	Drill Size	Threads Per Inch	Decimal Size	Drill Size	Threads Per Inch	Decimal Size
4"	3	4	3.75	–	–	–
3¾"	3	4	3.5	–	–	–
3½"	3	4	3.25			
3¼"	3	4	3.0	–	–	–
3"	2	4	2.75	–	–	–
2¾"	2	4	2.5	–	–	–
2½"	2	4	2.25	–	–	–
2¼"	2	4.5	2.0313	–	–	–
2"	1	4.5	1.7813	–	–	–
1¾"	1	2	1.5469	–	–	–
1½"	1	6	1.3281	1²⁷⁄₆₄"	12	1.4219
1⅜"	1	6	1.2188	1¹⁹⁄₆₄"	12	1.2969
1¼"	1	7	1.1094	1¹¹⁄₆₄"	12	1.1719
1⅛"	63⁄64"	7	.9844	1³⁄₆₄"	12	1.0469
1"	7⁄8"	8	.8750	15⁄16"	14	.9375
7⁄8"	49⁄64"	9	.7656	13⁄16"	14	.8125
3⁄4"	21⁄32"	10	.6563	11⁄16"	16	.6875
5⁄8"	17⁄32"	11	.5313	37⁄64"	18	.5781
9⁄16"	31⁄64"	12	.4844	33⁄64"	18	.5156
1⁄2"	27⁄64"	13	.4219	29⁄64"	20	.4531
7⁄16"	U	14	.368	25⁄64"	20	.3906
3⁄8"	5⁄16"	16	.3125	Q	24	.332
5⁄16"	F	18	.2570	I	24	.272
1⁄4"	#7	20	.201	#3	28	.213
#12	#16	24	.177	#14	28	.182
#10	#25	24	.1495	#21	32	.159
3⁄16"	#26	24	.147	#22	32	.157
#8	#29	32	.136	#29	36	.136
#6	#36	32	.1065	#33	40	.113
#5	#38	40	.1015	#37	44	.104
1⁄8"	3⁄32"	32	.0938	#38	40	.1015
#4	#43	40	.089	#42	48	.0935
#3	#47	48	.0785	#45	56	.082
#2	#50	56	.07	#50	64	.07
#1	#53	64	.0595	#53	72	.0595
#0	–	–	–	3⁄64"	80	.0469

MACHOS DE ROSCAR Y DADOS ESTÁNDAR (EN PULGADAS)

Tamaño de rosca	Grueso			Fine		
	Tamaño de broca	Roscas por pulgada	Tamaño decimal	Tamaño de broca	Rocas por pulgada	Tamaño decimal
4"	3	4	3.75	–	–	–
3¾"	3	4	3.5	–	–	–
3½"	3	4	3.25	–	–	–
3¼"	3	4	3.0	–	–	–
3"	2	4	2.75	–	–	–
2¾"	2	4	2.5	–	–	–
2½"	2	4	2.25	–	–	–
2¼"	2	4.5	2.0313	–	–	–
2"	1	4.5	1.7813	–	–	–
1¾"	1	2	1.5469	–	–	–
1½"	1	6	1.3281	1²⁷⁄₆₄"	12	1.4219
1⅜"	1	6	1.2188	1¹⁹⁄₆₄"	12	1.2969
1¼"	1	7	1.1094	1¹¹⁄₆₄"	12	1.1719
1⅛"	63⁄64"	7	.9844	1³⁄₆₄"	12	1.0469
1"	⅞"	8	.8750	15⁄16"	14	.9375
⅞"	49⁄64"	9	.7656	13⁄16"	14	.8125
¾"	21⁄32"	10	.6563	11⁄16"	16	.6875
⅝"	17⁄32"	11	.5313	37⁄64"	18	.5781
9⁄16"	31⁄64"	12	.4844	33⁄64"	18	.5156
½"	27⁄64"	13	.4219	29⁄64"	20	.4531
7⁄16"	U	14	.368	25⁄64"	20	.3906
⅜"	5⁄16"	16	.3125	Q	24	.332
5⁄16"	F	18	.2570	I	24	.272
¼"	#7	20	.201	#3	28	.213
#12	#16	24	.177	#14	28	.182
#10	#25	24	.1495	#21	32	.159
3⁄16"	#26	24	.147	#22	32	.157
#8	#29	32	.136	#29	36	.136
#6	#36	32	.1065	#33	40	.113
#5	#38	40	.1015	#37	44	.104
⅛"	3⁄32"	32	.0938	#38	40	.1015
#4	#43	40	.089	#42	48	.0935
#3	#47	48	.0785	#45	56	.082
#2	#50	56	.07	#50	64	.07
#1	#53	64	.0595	#53	72	.0595
#0	–	–	–	3⁄64"	80	.0469

TAPS AND DIES — METRIC CONVERSIONS

(mm) Thread Pitch	Fine Thread Size		Tap Drill Size	
	Inches	mm	Inches	mm
4.5	1.6535	42	1.4567	37.0
4.0	1.5748	40	1.4173	36.0
4.0	1.5354	39	1.3779	35.0
4.0	1.4961	38	1.3386	34.0
4.0	1.4173	36	1.2598	32.0
3.5	1.3386	34	1.2008	30.5
3.5	1.2992	33	1.1614	29.5
3.5	1.2598	32	1.1220	28.5
3.5	1.1811	30	1.0433	26.5
3.0	1.1024	28	.9842	25.0
3.0	1.0630	27	.9449	24.0
3.0	1.0236	26	.9055	23.0
3.0	.9449	24	.8268	21.0
2.5	.8771	22	.7677	19.5
2.5	.7974	20	.6890	17.5
2.5	.7087	18	.6102	15.5
2.0	.6299	16	.5118	14.0
2.0	.5512	14	.4724	12.0
1.75	.4624	12	.4134	10.5
1.50	.4624	12	.4134	10.5
1.50	.3937	11	.3780	9.6
1.50	.3937	10	.3386	8.6
1.25	.3543	9	.3071	7.8
1.25	.3150	8	.2677	6.8
1.0	.2856	7	.2362	6.0
1.0	.2362	6	.1968	5.0
.90	.2165	5.5	.1811	4.6
.80	.1968	5	.1653	4.2
.75	.1772	4.5	.1476	3.75
.70	.1575	4	.1299	3.3
.75	.1575	4	.1279	3.25
.60	.1378	3.5	.1142	2.9
.60	.1181	3	.0945	2.4
.50	.1181	3	.0984	2.5
.45	.1124	2.6	.0827	2.1
.45	.0984	2.5	.0787	2.0
.40	.0895	2.3	.0748	1.9
.40	.0787	2	.0630	1.6
.45	.0787	2	.0590	1.5
.35	.0590	1.5	.0433	1.1

Paso de rosca (mm)	MACHOS DE ROSCAR Y DADOS - CONVERSIONES A UNIDADES MÉTRICAS			
	Tamaño de rosca fina		Tamaño de broca	
	Pulgadas	mm	Pulgadas	mm
4.5	1.6535	42	1.4567	37.0
4.0	1.5748	40	1.4173	36.0
4.0	1.5354	39	1.3779	35.0
4.0	1.4961	38	1.3386	34.0
4.0	1.4173	36	1.2598	32.0
3.5	1.3386	34	1.2008	30.5
3.5	1.2992	33	1.1614	29.5
3.5	1.2598	32	1.1220	28.5
3.5	1.1811	30	1.0433	26.5
3.0	1.1024	28	.9842	25.0
3.0	1.0630	27	.9449	24.0
3.0	1.0236	26	.9055	23.0
3.0	.9449	24	.8268	21.0
2.5	.8771	22	.7677	19.5
2.5	.7974	20	.6890	17.5
2.5	.7087	18	.6102	15.5
2.0	.6299	16	.5118	14.0
2.0	.5512	14	.4724	12.0
1.75	.4624	12	.4134	10.5
1.50	.4624	12	.4134	10.5
1.50	.3937	11	.3780	9.6
1.50	.3937	10	.3386	8.6
1.25	.3543	9	.3071	7.8
1.25	.3150	8	.2677	6.8
1.0	.2856	7	.2362	6.0
1.0	.2362	6	.1968	5.0
.90	.2165	5.5	.1811	4.6
.80	.1968	5	.1653	4.2
.75	.1772	4.5	.1476	3.75
.70	.1575	4	.1299	3.3
.75	.1575	4	.1279	3.25
.60	.1378	3.5	.1142	2.9
.60	.1181	3	.0945	2.4
.50	.1181	3	.0984	2.5
.45	.1124	2.6	.0827	2.1
.45	.0984	2.5	.0787	2.0
.40	.0895	2.3	.0748	1.9
.40	.0787	2	.0630	1.6
.45	.0787	2	.0590	1.5
.35	.0590	1.5	.0433	1.1

RECOMMENDED DRILLING SPEEDS (RPMS)

Material	Bit Sizes	RPM Speed Range		
Glass	Special Metal Tube Drilling		700	
Plastics	7/16" and larger	500	–	1000
	3/8"	1500	–	2000
	5/16"	2000	–	2500
	1/4"	3000	–	3500
	3/16"	3500	–	4000
	1/8"	5000	–	6000
	1/16" and smaller	6000	–	6500
Woods	1" and larger	700	–	2000
	3/4" to 1"	2000	–	2300
	1/2" to 3/4"	2300	–	3100
	1/4" to 1/2"	3100	–	3800
	1/4" and smaller	3800	–	4000
	carving / routing	4000	–	6000
Soft Metals	7/16" and larger	1500	–	2500
	3/8"	3000	–	3500
	5/16"	3500	–	4000
	1/4"	4500	–	5000
	3/16"	5000	–	6000
	1/8"	6000	–	6500
	1/16" and smaller	6000	–	6500
Steel	7/16" and larger	500	–	1000
	3/8"	1000	–	1500
	5/16"	1000	–	1500
	1/4"	1500	–	2000
	3/16"	2000	–	2500
	1/8"	3000	–	4000
	1/16" and smaller	5000	–	6500
Cast Iron	7/16" and larger	1000	–	1500
	3/8"	1500	–	2000
	5/16"	1500	–	2000
	1/4"	2000	–	2500
	3/16"	2500	–	3000
	1/8"	3500	–	4500
	1/16" and smaller	6000	–	6500

VELOCIDADES RECOMENDADAS DE TALADRADO (RPMS)

Material	Tamaños de broca	Rango de velocidades en rpm		
Vidrio	Metal especial Taladrado tubular	700		
Plásticos	$^7/_{16}$" y mayores	500	–	1000
	$^3/_8$"	1500	–	2000
	$^5/_{16}$"	2000	–	2500
	$^1/_4$"	3000	–	3500
	$^3/_{16}$"	3500	–	4000
	$^1/_8$"	5000	–	6000
	$^1/_{16}$" y menores	6000	–	6500
Maderas	1" y mayores	700	–	2000
	$^3/_4$" a 1"	2000	–	2300
	$^1/_2$" a $^3/_4$"	2300	–	3100
	$^1/_4$" a $^1/_2$"	3100	–	3800
	$^1/_4$" y menores	3800	–	4000
	tallado / fresado	4000	–	6000
Metales blandos	$^7/_{16}$" y mayores	1500	–	2500
	$^3/_8$"	3000	–	3500
	$^5/_{16}$"	3500	–	4000
	$^1/_4$"	4500	–	5000
	$^3/_{16}$"	5000	–	6000
	$^1/_8$"	6000	–	6500
	$^1/_{16}$" y menores	6000	–	6500
Acero	$^7/_{16}$" y mayores	500	–	1000
	$^3/_8$"	1000	–	1500
	$^5/_{16}$"	1000	–	1500
	$^1/_4$"	1500	–	2000
	$^3/_{16}$"	2000	–	2500
	$^1/_8$"	3000	–	4000
	$^1/_{16}$" y menores	5000	–	6500
Hierro fundido	$^7/_{16}$" y mayores	1000	–	1500
	$^3/_8$"	1500	–	2000
	$^5/_{16}$"	1500	–	2000
	$^1/_4$"	2000	–	2500
	$^3/_{16}$"	2500	–	3000
	$^1/_8$"	3500	–	4500
	$^1/_{16}$" y menores	6000	–	6500

TORQUE LUBRICATION EFFECTS IN FOOT-POUNDS

Lubricant	5/16" - 18 Thread	1/2" - 13 Thread	Torque Decrease
Graphite	13	62	49 - 55%
Mily Film	14	66	45 - 52%
White Grease	16	79	35 - 45%
Sae 30	16	79	35 - 45%
Sae 40	17	83	31 - 41%
Sae 20	18	87	28 - 38%
Plated	19	90	26 - 34%
No Lube	29	121	0%

METALWORKING LUBRICANTS

Materials	Threading	Lathing	Drilling
Machine Steels	Dissolvable Oil Mineral Oil Lard Oil	Dissolvable Oil	Dissolvable Oil Sulpherized Oil Min. Lard Oil
Tool Steels	Lard Oil Sulpherized Oil	Dissolvable Oil	Dissolvable Oil Sulpherized Oil
Cast Irons	Sulpherized Oil Dry Min. Lard Oil	Dissolvable Oil Dry	Dissolvable Oil Dry Air Jet
Malleable Irons	Soda Water Lard Oil	Soda Water Dissolvable Oil	Soda Water Dry
Aluminums	Kerosene Dissolvable Oil Lard Oil	Dissolvable Oil	Kerosene Dissolvable Oil
Brasses	Dissolvable Oil Lard Oil	Dissolvable Oil	Kerosene Dissolvable Oil Dry
Bronzes	Dissolvable Oil Lard Oil	Dissolvable Oil	Dissolvable Oil Dry
Coppers	Dissolvable Oil Lard Oil	Dissolvable Oil	Kerosene Dissolvable Oil Dry

EFECTOS DE LA TORSIÓN EN LA LUBRICACIÓN EN PIES – LIBRA			
Lubricante	Rosca 5/16" - 18	Rosca 1/2" - 13	Disminución de torsión
Grafito	13	62	49 - 55%
Película Mily	14	66	45 - 52%
Grasa blanca	16	79	35 - 45%
Sae 30	16	79	35 - 45%
Sae 40	17	83	31 - 41%
Sae 20	18	87	28 - 38%
Revestido	19	90	26 - 34%
Sin lubricación	29	121	0%

LUBRICANTES PARA METALMECÁNICA

Materiales	Roscado	Listonado	Taladrado
Acero de máquinas	Aceite soluble Aceite mineral Aceite de grasa de cerdo	Aceite soluble	Aceite soluble Aceite sulfurado Aceite de grasa de cerdo mineral
Acero para herramientas	Aceite de grasa de cerdo Aceite sulfurado	Aceite soluble	Aceite soluble Aceite sulfurado
Hierros fundidos	Aceite sulfurado Seco Aceite de grasa de cerdo mineral	Aceite soluble Seco	Aceite soluble Seco Chorro de aire a presión
Hierros maleables	Soda (agua gasificada) Aceite de grasa de cerdo	Soda (agua gasificada) Aceite soluble	Soda (agua gasificada) Seco
Aluminios	Queroseno Aceite soluble Aceite de grasa de cerdo	Aceite soluble	Queroseno
Bronces	Aceite soluble Aceite de grasa de cerdo	Aceite soluble	Queroseno Aceite soluble Seco
Bronces	Aceite soluble Aceite de grasa de cerdo	Aceite soluble	Aceite soluble Seco
Cobres	Aceite soluble Aceite de grasa de cerdo	Aceite soluble	Queroseno Aceite soluble Seco

TYPES OF SOLDERING FLUX

To Solder	Use
For cast iron	Cuprous oxide
For galvanized iron, galvanized, steel, tin, zinc	Hydrochloric acid
For pewter and lead	Organic
For brass, copper, gold, iron, silver, steel	Borax
For brass, bronze, cadmium, copper, lead, silver	Resin
For brass, copper, gun metal, iron, nickel, tin, zinc	Ammonia chloride
For bismuth, brass, copper, gold, silver, tin	Zinc chloride
For silver	Sterling
For pewter and lead	Tallow
For stainless only	Stainless steel (only)

HARD SOLDER ALLOYS

To hard solder	Copper %	Gold %	Silver %	Zinc %
Gold	22	67	11	–
Silver	20	–	70	10
Hard brass	45	–	–	55
Soft brass	22	–	–	78
Copper	50	–	–	50
Cast iron	55	–	–	45
Steel and iron	64	–	–	36

SOFT SOLDER ALLOYS

To soft solder	Lead %	Tin %	Zinc %	Bism %	Other %
Gold	33	67	–	–	–
Silver	33	67	–	–	–
Brass	34	66	–	–	–
Copper	40	60	–	–	–
Steel and iron	50	50	–	–	–
Galvanized steel	42	58	–	–	–
Tinned steel	36	64	–	–	–
Zinc	45	55	–	–	–
Block Tin	1	99	–	–	–
Lead	67	33	–	–	–
Gun metal	37	63	–	–	–
Pewter	25	25	–	50	–
Bismuth	33	33	–	34	–
Aluminum	–	70	25	–	5

TIPOS DE SOLDADURA

Para soldar	Use
Para hierro fundido	Óxido cuproso
Para hierro galvanizado, acero galvanizado, estaño y zinc	Ácido hidroclórico
Para peltre y plomo	Material orgánico
Para bronce, cobre, oro, hierro, plata y acero	Bórax
Para bronce, bronce, cadmio, cobre, plomo y plata	Resina
Para bronce, cobre, bronce de cañón, hierro, níquel, estaño y zinc	Cloruro de amonio
Para bismuto, bronce, cobre, oro, plata y estaño	Cloruro de zinc
Para plata	Plata de ley
Para peltre y plomo	Sebo
Para acero inoxidable solamente	Acero inoxidable (solamente)

ALEACIONES DURAS PARA SOLDAR

Para soldar	Cobre %	Oro %	Plata %	Zinc %
Oro	22	67	11	–
Plata	20	–	70	10
Bronce duro	45	–	–	55
Bronce blando	22	–	–	78
Cobre	50	–	–	50
Hierro fundido	55	–	–	45
Acero y hierro	64	–	–	36

ALEACIONES PARA SOLDADURAS BLANDAS

Para soldar con	Plomo %	Estaño %	Zinc %	Bismuto %	Otros %
Oro	33	67	–	–	–
Plata	33	67	–	–	–
Bronce	34	66	–	–	–
Cobre	40	60	–	–	–
Acero y hierro	50	50	–	–	–
Acero galvanizado	42	58	–	–	–
Acero estañado	36	64	–	–	–
Zinc	45	55	–	–	–
Estaño en lingotes	1	99	–	–	–
Plomo	67	33	–	–	–
Bronce de cañón	37	63	–	–	–
Peltre	25	25	–	50	–
Bismuto	33	33	–	34	–
Aluminio	–	70	25	–	5

PROPERTIES OF WELDING GASES

Type of Gas	Characteristics	Common Tank Sizes (cu. ft.)
Acetylene	C_2H_2, explosive gas, flammable, garlic - like odor, colorless, dangerous if used in pressures over 15 psig (30 psig absolute)	10, 40, 75 100, 300
Argon	Ar, non-explosive inert gas, tasteless, odorless, colorless	131, 330 4754 (Liquid)
Carbon Dioxide	CO_2, Non-explosive inert gas, tasteless, odorless, colorless (in large quantities is toxic)	20 lbs., 50 lbs.
Helium	He, Non-explosive inert gas, tasteless, odorless, colorless	221
Hydrogen	H2, explosive gas, tasteless, odorless, colorless	191
Nitrogen	N2, Non-explosive inert gas, tasteless, odorless, colorless	20, 40, 80 113, 225
Oxygen	O2, Non-explosive gas, tasteless, odorless, colorless, supports combustion	20, 40, 80 122, 244 4500 (liquid)

WELDING RODS – 36" LONG

Rod (Dia.) Size	Number of Rods Per Pound			
	Aluminum	Brass	Cast Iron	Steel
3/8"	–	1.0	.25	1.0
5/16"	–	–	.50	1.33
1/4"	6.0	2.0	2.25	2.0
3/16"	9.0	3.0	5.50	3.5
5/32"	–	–	–	5.0
1/8"	23.0	7.0	–	8.0
3/32"	41.0	13.0	–	14.0
1/16"	91.0	29.0	–	31.0

PROPIEDADES DE LOS GASES PARA SOLDAR

Tipo de Gas	Características	Tamaños comunes de tanques (pies cúbicos)
Acetileno	C_2H_2, gas explosivo, inflamable, de olor similar al del ajo, incoloro, peligroso si se usa a presiones superiores a 15 psig (30 psig absolutos)	10, 40, 75 100, 300
Argón	Ar, gas inerte no explosivo, insípido, inodoro, incoloro	131, 330 4754 (Líquido)
Dióxido de carbono	CO_2, gas inerte no explosivo, insípido, inodoro, incoloro (en grandes cantidades es tóxico)	20 lbs, 50 lbs
Helio	He, gas inerte no explosivo, insípido, inodoro, incoloro	221
Hidrógeno	H2, gas explosivo, insípido, inodoro, incoloro	191
Nitrógeno	N2, gas inerte no explosivo, insípido, inodoro, incoloro	20, 40, 80 113, 225
Oxígeno	O2, gas no explosivo, insípido, inodoro, incoloro, combustible	20, 40, 80 122, 244 4500 (líquido)

VARILLAS PARA SOLDAR – 36" DE LONGITUD

Tamaño (Diám.) de varilla	Número de varillas por libra			
	Aluminio	Bronce	Hierro fundido	Acero
3/8"	–	1.0	.25	1.0
5/16"	–	–	.50	1.33
1/4"	6.0	2.0	2.25	2.0
3/16"	9.0	3.0	5.50	3.5
5/32"	–	–	–	5.0
1/8"	23.0	7.0	–	8.0
3/32"	41.0	13.0	–	14.0
1/16"	91.0	29.0	–	31.0

CABLE CLAMPS PER WIRE ROPE SIZE

Wire Rope Diameter (Inches)	# of Clamps Required	Clip Spacing (Inches)	Rope Turn-back (Inches)
1/8	2	3	3 1/4
3/16	2	3	3 3/4
1/4	2	3 1/4	4 3/4
5/16	2	3 1/4	5 1/4
3/8	2	4	6 1/2
7/16	2	4 1/2	4
1/2	3	5	11 1/2
9/16	3	5 1/2	12
5/8	3	5 3/4	12
3/4	4	6 3/4	18
7/8	4	8	19
1	5	8 3/4	26
1 1/8	6	9 3/4	34
1 1/4	6	10 3/4	37
1 7/16	7	11 1/2	44
1 1/2	7	12 1/2	48
1 5/8	7	13 1/4	51
1 3/4	7	14 1/2	53
2	8	16 1/2	71
2 1/4	8	16 1/2	73
2 1/2	9	17 3/4	84
2 3/4	10	18	100
3	10	18	106

ABRAZADERAS DE CABLES POR TAMAÑO DE CABLE

Diámetro del cable de alambre (pulgadas)	Núm. de abrazaderas requeridas	Espaciamiento entre broches (pulgadas)	Retorno de cable (pulgadas)
1/8	2	3	3 1/4
3/16	2	3	3 3/4
1/4	2	3 1/4	4 3/4
5/16	2	3 1/4	5 1/4
3/8	2	4	6 1/2
7/16	2	4 1/2	4
1/2	3	5	11 1/2
9/16	3	5 1/2	12
5/8	3	5 3/4	12
3/4	4	6 3/4	18
7/8	4	8	19
1	5	8 3/4	26
1 1/8	6	9 3/4	34
1 1/4	6	10 3/4	37
1 7/16	7	11 1/2	44
1 1/2	7	12 1/2	48
1 5/8	7	13 1/4	51
1 3/4	7	14 1/2	53
2	8	16 1/2	71
2 1/4	8	16 1/2	73
2 1/2	9	17 3/4	84
2 3/4	10	18	100
3	10	18	106

TYPES OF FIRE EXTINGUISHERS

Today they are virtually standard equipment in a business or residence and are rated by the make up of the fire they will extinguish.

TYPE A: To extinguish fires involving trash, cloth, paper and other wood or pulp based materials. The flames are put out by water based ingredients or dry chemicals.

TYPE B: To extinguish fires involving greases, paints, solvents, gas and other petroleum based liquids. The flames are put out by cutting off oxygen and stopping the release of flammable vapors. Dry chemicals, foams and halon are used.

TYPE C: To extinguish fires involving electricity. The combustion is put out the same way as with a type B extinguisher, but, most importantly, the chemical in a type C <u>MUST</u> be non-conductive to electricity in order to be safe and effective.

TYPE D: To extinguish fires involving combustible metals. Please be advised to obtain important information from your local fire department on the requirements for type D fire extinguishers for your area.

Any combination of letters indicate that an extinguisher will put out more than one type of fire. A type BC will put out two types of fires. The size of the fire to be extinguished is shown by a number in front of the letter such as 100A.
The following formulas apply:

Class "1A": Will extinguish 25 burning sticks 40 inches long.

Class "1B": Will extinguish a paint thinner fire 2.5 square feet in size.

A 100B fire extinguisher will put out a fire 100 times larger than a type 1B.

Here are some basic guidelines to follow:

- By using a type ABC you will cover most basic fires.
- Use fire extinguishers with a gauge and ones that are constructed with metal. Also note if the unit is U.L. approved.
- Utilize more than one extinguisher and be sure that each unit is mounted in a clearly visible and accessible manner.
- After purchasing any fire extinguisher always review the basic instructions for its intended use. Never deviate from the manufacturers' guidelines. Following this simple procedure could end up saving lives.

TIPOS DE EXTINGUIDORES DE INCENDIO

Hoy en día son prácticamente equipos normales en empresas o residencias y se clasifican por el tipo de incendio que extinguen.

Tipo A: Para extinguir incendios ocasionados por desechos, tela, papel y otros materiales basados en madera o pulpa. Las llamas se neutralizan por ingredientes basados en agua o sustancias químicas secas.

Tipo B: Para extinguir incendios relacionados con grasas, pinturas, solventes, gas y otros líquidos derivados del petróleo. Las llamas se neutralizan eliminando el oxígeno y deteniendo la liberación de vapores inflamables. Se usan sustancias químicas secas, espuma y halón.

Tipo C: Para extinguir incendios causados por desperfectos eléctricos. La combustión se neutraliza de la misma manera que con un extinguidor tipo B, pero, muy importante, el agente químico en un equipo de tipo C **DEBE** ser no conductivo de la electricidad para ser seguro y eficaz.

Tipo D: Para extinguir incendios vinculados con metales combustibles. Procure obtener información importante de su departamento de bomberos sobre los requisitos para los extinguidores de incendio tipo D en su localidad.

Cualquier combinación de letras indicará que un extinguidor neutralizará más de un tipo de incendio. Un tipo BC neutralizará dos tipos de incendio. El tamaño del incendio a extinguir se muestra mediante un número antes de la letra, como 100A. Rigen las siguientes fórmulas:.

Clase "1A": Extinguirá 25 varillas ardientes de 40 pulgadas de largo cada una.

Clase "1B": Extinguirá un incendio causado por diluyentes de pintura en un área de 2.5 pies cuadrados de tamaño

Un extinguidor de incendios 100B neutralizará un incendio 100 veces más grande que uno tipo 1B

He aquí algunas pautas básicas a seguir:

- Con el empleo de un extinguidor tipo ABC se cubren la mayoría de los incendios básicos.
- Utilice extinguidores de incendio con manómetro y que estén construidos de metal. También verifique que U.L. apruebe el equipo.
- Utilice más de un extinguidor y asegúrese de que cada equipo esté montado de manera de ser claramente visible y accesible.
- Después de comprar un extinguidor de incendios siempre examine sus instrucciones básicas de empleo. Nunca se aparte de las pautas provistas pro el fabricante. La aplicación de estos sencillos procedimientos puede terminar salvando vidas.

PULLEY AND GEAR FORMULAS

For single reduction or increase of speed by means of belting where the speed at which each shaft should run is known, and one pulley is in place:

Multiply the diameter of the pulley which you have by the number of revolutions per minute that its shaft makes; divide this product by the speed in revolutions per minute at which the second shaft should run. The result is the diameter of pulley to use.

Where both shafts with pulleys are in operation and the speed of one is known:

Multiply the speed of the shaft by diameter of its pulley and divide this product by diameter of pulley on the other shaft. The result is the speed of the second shaft.

Where a countershaft is used, to obtain size of main driving or driven pulley, or speed of main driving or driven shaft, it is necessary to calculate, as above, between the known end of the transmission and the countershaft, then repeat this calculation between the countershaft and the unknown end.

A set of gears of the same pitch transmits speeds in proportion to the number of teeth they contain. Count the number of teeth in the gear wheel and use this quantity instead of the diameter of pulley, mentioned above, to obtain number of teeth cut in unknown gear, or speed of second shaft.

Formulas For Finding Pulley Sizes:

$$d = \frac{D \times S}{s'} \qquad D = \frac{d \times s'}{S}$$

d = diameter of driven pulley

D = diameter of driving pulley

s' = number of revolutions per minute of driven pulley

S = number of revolutions per minute of driving pulley

CÁLCULOS CON POLEAS Y ENGRANAJES

Para reducción o incremento individual de la velocidad por medio de correas cuando la velocidad a la que cada eje deberá operar se conoce y se dispone de una polea:

Multiplique el diámetro de la polea de la que dispone por el número de revoluciones por minuto que realiza su eje; divida ese producto por la velocidad en revoluciones por minuto a la cual deberá operar el segundo eje. El resultado es el diámetro de la polea a usar.

Cuando ambos ejes con sus poleas están en operación y la velocidad de uno de ellos se conoce:

Multiplique la velocidad del eje por el diámetro de su polea y divida ese producto por el diámetro de la polea del otro eje. El resultado es la velocidad del segundo eje.

Cuando se usa un contraeje, para obtener el tamaño de la polea motriz o impulsada principal, o la velocidad del eje motriz principal o impulsado, es necesario proceder a calcular, como más arriba, entre el extremo conocido de la transmisión y el contraeje, y luego repetir dicho cálculo entre el contraeje y el extremo desconocido.

Un juego de engranajes de la misma separación transmite velocidades en proporción al número de dientes que contienen los componentes del mismo. Cuente el número de dientes en la rueda de engranajes y use dicha cantidad en lugar del diámetro de la polea, mencionado más arriba, para obtener el número de dientes del engranaje desconocido o la velocidad del segundo eje.

Fórmulas Para Determinar Tamaños De Poleas

$$d = \frac{D \times S}{s'} \qquad D = \frac{d \times s'}{S}$$

d = diámetro de la polea impulsada.

D = diámetro de la polea motriz.

S = número de revoluciones por minuto de la polea motriz.

s' = número de revoluciones por minuto de la polea impulsada.

PULLEY AND GEAR FORMULAS (cont.)

Formulas For Finding Gear Sizes:

$$n = \frac{N \times S}{s'} \qquad N = \frac{n \times s'}{S}$$

n = number of teeth in pinion (driving gear)

N = number of teeth in gear (driven gear)

s' = number of revolutions per minute of gear

S = number of revolutions per minute of pinion

Formula To Determine Shaft Diameter:

$$\text{diameter of shaft in inches} = \sqrt[3]{\frac{K \times HP}{RPM}}$$

HP = the horsepower to be transmitted

RPM = speed of shaft

K = factor which varies from 50 to 125 depending on type of shaft and distance between supporting bearings.

For line shaft having bearings 8 feet apart:

K = 90 for turned shafting

K = 70 for cold-rolled shafting

Formula To Determine Belt Length:

$$\text{length of belt} = \frac{3.14\,(D + d)}{2} + 2\left(\sqrt{X^2 + \left(\frac{D - d}{2}\right)^2}\right)$$

D = diameter of large pulley

d = diameter of small pulley

X = distance between centers of shafting

CÁLCULOS CON POLEAS Y ENGRANAJES (cont.)

Fórmulas para determinar tamaños de engranajes:

$$n = \frac{N \times S}{s'} \qquad N = \frac{n \times s'}{S}$$

n = número de dientes del piñón (engranaje motriz).

N = número de dientes del engranaje (engranaje impulsado).

S = número de revoluciones por minuto del piñón.

s' = número de revoluciones por minuto del engranaje.

Fórmulas para determinar el diámetro del eje:

La fórmula para determinar el tamaño del eje de acero para transmitir una potencia dada a una velocidad determinada es la siguiente:

$$\text{diámetro del eje en pulgadas} = \sqrt[3]{\frac{K \times HP}{RPM}}$$

donde HP = la potencia en HP que se transmitirá

RPM = velocidad del eje

K = factor que varía entre 50 y 125 según el tipo de eje y la distancia entre los rodamientos de apoyo.

Para árboles de transmisión con rodamientos espaciados 8 pies entre sí:

K = 90 para ejes laminados en caliente

K = 70 para ejes laminados en frío

Fórmula para determinar la longitud de la correa:

Para determinar la longitud de la correa se usa la siguiente fórmula:

$$\text{longitud de la correa} = \frac{3.14\,(D+d)}{2} + 2\left(\sqrt{X^2 + \left(\frac{D-d}{2}\right)^2}\right)$$

donde D = diámetro de la polea grande

d = diámetro de la polea pequeña

X = distancia entre los centros de los ejes

STANDARD "V" BELT LENGTHS IN INCHES

A BELTS

Standard Belt No.	Pitch Length	Outside Length
A26	27.3	28.0
A31	32.3	33.0
A35	36.3	37.0
A38	39.3	40.0
A42	43.3	44.0
A46	47.3	48.0
A51	52.3	53.0
A55	56.3	57.0
A60	61.3	62.0
A68	69.3	70.0
A75	76.3	77.0
A80	81.3	82.0
A85	86.3	87.0
A90	91.3	92.0
A96	97.3	98.0
A105	106.3	107.0
A112	113.3	114.0
A120	121.3	122.0
A128	129.3	130.0

B BELTS

Standard Belt No.	Pitch Length	Outside Length
B35	36.8	38.0
B38	39.8	41.0
B42	43.8	45.0
B46	47.8	49.0
B51	52.8	54.0
B55	56.8	58.0
B60	61.8	63.0
B68	69.8	71.0
B75	76.8	78.0
B81	82.8	84.0
B85	86.8	88.0
B90	91.8	93.0
B97	98.8	100.0
B105	106.8	108.0
B112	113.8	115.0
B120	121.8	123.0
B128	129.8	131.0
B136	137.8	139.0
B144	145.8	147.0
B158	159.8	161.0
B173	174.8	176.0
B180	181.8	183.0
B195	196.8	198.0
B210	211.8	213.0
B240	240.3	241.5
B270	270.3	271.5
B300	300.3	301.5

C BELTS

Standard Belt No.	Pitch Length	Outside Length
C51	53.9	55.0
C60	62.9	64.0
C68	70.9	72.0
C75	77.9	79.0
C81	83.9	85.0
C85	87.9	89.0
C90	92.9	94.0
C96	98.9	100.0
C105	107.9	109.0
C112	114.9	116.0
C120	122.9	124.0
C128	130.9	132.0
C136	138.9	140.0
C144	146.9	148.0
C158	160.9	162.0
C162	164.9	166.0
C173	175.9	177.0
C180	182.9	184.0
C195	197.9	199.0
C210	212.9	214.0
C240	240.9	242.0
C270	270.9	272.0
C300	300.9	302.0
C360	360.9	362.0
C390	390.9	392.0
C420	420.9	422.0

D BELTS

Standard Belt No.	Pitch Length	Outside Length
D128	131.3	133.0
D144	147.3	149.0
D158	161.3	163.0
D162	165.3	167.0
D173	176.3	178.0
D180	183.3	185.0
D195	198.3	200.0
D210	213.3	215.0
D240	240.8	242.0
D270	270.8	272.5
D300	300.8	302.5
D330	330.8	332.5
D360	360.8	362.5
D390	390.8	392.5
D420	420.8	422.5
D480	480.8	482.5
D540	540.8	542.5
D600	600.8	602.5

E BELTS

Standard Belt No.	Pitch Length	Outside Length	Standard Belt No.	Pitch Length	Outside Length
E180	184.5	187.5	E360	361.0	364.0
E195	199.5	202.5	E390	391.0	394.0
E210	214.5	217.5	E420	421.0	424.0
E240	241.0	244.0	E480	481.0	484.0
E270	271.0	274.0	E540	541.0	544.0
E300	301.0	304.0	E600	601.0	604.0
E330	331.0	334.0			

LONGITUDES NORMALES DE CORREAS EN "V"

CORREAS A			CORREAS B			CORREAS C		
Núm. de CORREA normal	Longitud del espaciado	Longitud exterior	Núm. de CORREA normal	Longitud del espaciado	Longitud exterior	Núm. de CORREA normal	Longitud del espaciado	Longitud exterior
A26	27.3	28.0	B35	36.8	38.0	C51	53.9	55.0
A31	32.3	33.0	B38	39.8	41.0	C60	62.9	64.0
A35	36.3	37.0	B42	43.8	45.0	C68	70.9	81.0
A38	39.3	40.0	B46	47.8	49.0	C75	77.9	79.0
A42	43.3	44.0	B51	52.8	54.0	C81	83.9	85.0
A46	47.3	48.0	B55	56.8	58.0	C85	87.9	89.0
A51	52.3	53.0	B60	61.8	63.0	C90	92.9	94.0
A55	56.3	57.0	B68	69.8	71.0	C96	98.9	100.0
A60	61.3	62.0	B75	76.8	78.0	C105	107.9	109.0
A68	69.3	70.0	B81	82.8	84.0	C112	114.9	116.0
A75	76.3	77.0	B85	86.8	88.0	C120	122.9	124.0
A80	81.3	82.0	B90	91.8	93.0	C128	130.9	132.0
A85	86.3	87.0	B97	98.8	100.0	C136	138.9	140.0
A90	91.3	92.0	B105	106.8	108.0	C144	146.9	148.0
A96	97.3	98.0	B112	113.8	115.0	C158	160.9	162.0
A105	106.3	107.0	B120	121.8	123.0	C162	164.9	166.0
A112	113.3	114.0	B128	129.8	131.0	C173	175.9	177.0
A120	121.3	122.0	B136	137.8	139.0	C180	182.9	184.0
A128	129.3	130.0	B144	145.8	147.0	C195	197.9	199.0
			B158	159.8	161.0	C210	212.9	214.0
			B173	174.8	176.0	C240	240.9	242.0

CORREAS D		
Núm. de CORREA normal	Longitud del espaciado	Longitud exterior
D128	131.3	133.0
D144	147.3	149.0
D158	1613	163.0
D162	165.3	167.0
D173	176.3	178.0
D180	183.3	185.0
D195	198.3	200.0
D210	213.3	215.0
D240	240.8	242.0
D270	270.8	272.5
D300	300.8	302.5
D330	330.8	332.5
D360	360.8	362.5
D390	390.8	392.5
D420	420.8	422.5
D480	480.8	482.5
D540	540.8	542.5
D600	600.8	602.5

(CORREAS B continuación)

Núm. de CORREA normal	Longitud del espaciado	Longitud exterior
B180	181.8	183.0
B195	196.8	198.0
B210	211.8	213.0
B240	240.3	241.5
B270	270.3	271.5
B300	300.3	301.5

(CORREAS C continuación)

Núm. de CORREA normal	Longitud del espaciado	Longitud exterior
C270	270.9	272.0
C300	300.9	302.0
C360	360.9	362.0
C390	390.9	392.0
C420	420.9	422.0

CORREAS E			CORREAS E		
Núm. de CORREA normal	Longitud del espaciado	Longitud exterior	Núm. de CORREA normal	Longitud del espaciado	Longitud exterior
E180	184.5	187.5	E360	361.0	364.0
E195	199.5	202.5	E390	391.0	394.0
E210	214.5	217.5	E420	421.0	424.0
E240	241.0	244.0	E480	481.0	484.0
E270	271.0	274.0	E540	541.0	544.0
E300	301.0	304.0	E600	601.0	604.0
E330	331.0	334.0			

STANDARD "V" BELT LENGTHS IN INCHES (cont.)

3V Belts		5V Belts		8V Belts	
3V250	25	5V500	50	8V1000	100
3V265	26.5	5V530	53	8V1060	106
3V280	28	5V560	56	8V1120	112
3V300	30	5V600	60	8V1180	118
3V315	31.5	5V630	63	8V1250	125
3V335	33.5	5V670	67	8V1320	132
3V355	35.5	5V710	71	8V1400	140
3V375	37.5	5V750	75	8V1500	150
3V400	40	5V800	80	8V1600	160
3V425	42.5	5V850	85	8V1700	170
3V450	45	5V900	90	8V1800	180
3V475	47.5	5V950	95	8V1900	190
3V500	50	5V1000	100	8V2000	200
3V530	53	5V1060	106	8V2120	212
3V560	56	5V1120	112	8V2240	224
3V600	60	5V1180	118	8V2360	236
3V630	63	5V1250	125	8V2500	250
3V670	67	5V1320	132	8V2650	265
3V710	71	5V1400	140	8V2800	280
3V750	75	5V1500	150	8V3000	300
3V800	80	5V1600	160	8V3150	315
3V850	85	5V1700	170	8V3350	335
3V900	90	5V1800	180	8V3550	355
3V950	95	5V1900	190	8V3750	375
3V1000	100	5V2000	200	8V4000	400
3V1060	106	5V2120	212	8V4250	425
3V1120	112	5V2240	224	8V4500	450
3V1180	118	5V2360	236	8V5000	500
3V1250	128	5V2500	250		
3V1320	132	5V2650	265		
3V1400	140	5V2800	280		
		5V3000	300		
		5V3150	315		
		5V3350	335		
		5V3550	355		

If the 60-inch "B" section belt shown is made 3/10 of an inch longer, it will be code marked 53 rather than 50. If made 3/10 shorter, it will be marked 47. While both have the belt number B60 they cannot be used in a set because of the difference in length.

TYPICAL CODE MARKING

B60   MANUFACTURER'S NAME   50

NOMINAL
SIZE AND LENGTH

LENGTH
CODE NUMBER

LONGITUDES NORMALES DE CORREAS EN "V" (cont.)

CORREAS 3V		CORREAS 5V		CORREAS 8V	
3V250	25	5V500	50	8V1000	100
3V265	26.5	5V530	53	8V1060	106
3V280	28	5V560	56	8V1120	112
3V300	30	5V600	60	8V1180	118
3V315	31.5	5V630	63	8V1250	125
3V335	33.5	5V670	67	8V1320	132
3V355	35.5	5V710	71	8V1400	140
3V375	37.5	5V750	75	8V1500	150
3V400	40	5V800	80	8V1600	160
3V425	42.5	5V850	85	8V1700	170
3V450	45	5V900	90	8V1800	180
3V475	47.5	5V950	95	8V1900	190
3V500	50	5V1000	100	8V2000	200
3V530	53	5V1060	106	8V2120	212
3V560	56	5V1120	112	8V2240	224
3V600	60	5V1180	118	8V2360	236
3V630	63	5V1250	125	8V2500	250
3V670	67	5V1320	132	8V2650	265
3V710	71	5V1400	140	8V2800	280
3V750	75	5V1500	150	8V3000	300
3V800	80	5V1600	160	8V3150	315
3V850	85	5V1700	170	8V3350	335
3V900	90	5V1800	180	8V3550	355
3V950	95	5V1900	190	8V3750	375
3V1000	100	5V2000	200	8V4000	400
3V1060	106	5V2120	212	8V4250	425
3V1120	112	5V2240	224	8V4500	450
3V1180	118	5V2360	236	8V5000	500
3V1250	128	5V2500	250		
3V1320	132	5V2650	265		
3V1400	140	5V2800	280		
		5V3000	300		
		5V3150	315		
		5V3350	335		
		5V3550	355		

Por ejemplo, si la correa de sección "B" de 60 pulgadas que se muestra se fabrica 3/10 de pulgada más larga, se marcará con el código 53 en lugar del 50 mostrado. Si se fabricara 3/10 de pulgada más corta, se marcaría con el código 47. Aunque ambas correas tienen el número de correa B60, no pueden usarse satisfactoriamente en un juego debido a la diferencia en sus longitudes reales.

MARCADO TÍPICO DEL CÓDIGO

| 860 | NOMBRE DEL FABRICANTE | 50 |

LONGITUD Y TAMAÑO NOMINALES   NÚMERO DE CÓDIGO DE LONGITUD

MELTING POINT AND RELATIVE CONDUCTIVITY OF DIFFERENT METALS AND ALLOYS

Metals	Relative Conductivity	Melting Point °F
Pure silver	106.0	1760
Pure copper	100.0	1980
Refined and crystalized copper	99.9	—
Telegraphic silicious bronze	98.0	—
Alloy of copper and silver (50%)	86.65	—
Silicide of copper, 4% Si	75.0	—
Pure gold	71.3	1950
Pure aluminum	64.5	1220
Silicide of copper, 12% Si	54.7	—
Tin with 12% of sodium	46.9	—
Telephonic silicious bronze	35.0	—
Copper with 10% of lead	30.0	—
Pure zinc	29.2	790
Telephonic phosphor-bronze	29.0	—
Silicious brass, 25% zinc	26.4	—
Brass with 35% zinc	21.59	—
Phosphor-tin	17.7	—
Alloy of gold and silver (50%)	16.12	—
Swedish iron	16.4	2800
Pure platinum	16.3	3230
Pure banca tin	15.5	442
Antimonial copper	12.7	—
Aluminum bronze (10%)	12.6	—
Siemens Steel	12.0	—
Copper with 10% of nickel	10.6	—
Cadmium amalgam (15%)	10.2	—
Dronier mercurial bronze	10.14	—
Arsenical copper (10%)	9.1	—
Bronze with 20% tin	8.4	—
Pure lead	7.8	620
Phosphor-bronze, 10% tin	6.5	—
Pure nickel	5.8	2640
Phosphor-copper, 9% phosphor	4.9	—
Antimony	4.4	1167

PUNTO DE FUSIÓN Y CONDUCTIVIDAD RELATIVA DE DIFERENTES METALES Y ALEACIONES

Metales	Conductividad relativa	Punto de fusión °F
Plata pura	106.0	1760
Cobre puro	100.0	1980
Cobre refinado y cristalizado	99.9	——
Bronce al silicio de grado telegráfico	98.0	——
Aleación de cobre y plata (50%)	86.65	——
Siliciuro de cobre, 4% Si	75.0	——
Oro puro	71.3	1950
Aluminio puro	64.5	1220
Siliciuro de cobre, 12% Si	54.7	——
Estaño con 12% de sodio	46.9	——
Bronce al silicio de grado telefónico	35.0	——
Cobre con 10% de plomo	30.0	——
Zinc puro	29.2	790
Fósforo-bronce de grado telefónico	29.0	——
Bronce al silicio, 25% de zinc	26.4	——
Bronce con 35% de zinc	21.59	——
Fósforo-estaño	17.7	——
Aleación de oro y plata (50%)	16.12	——
Hierro sueco	16.4	2800
Platino puro	16.3	3230
Estaño banca puro	15.5	442
Cobre con antimonio	12.7	——
Bronce con aluminio (10%)	12.6	——
Acero Siemens	12.0	——
Cobre con 10% de níquel	10.6	——
Amalgama de cadmio (15%)	10.2	——
Bronce mercuriado Dronier	10.14	——
Cobre arsenical (10%)	9.1	——
Bronce con 20% de estaño	8.4	——
Plomo puro	7.8	620
Fósforo-bronce, 10% de estaño	6.5	——
Níquel puro	5.8	2640
Fósforo-cobre, 9% de fósforo	4.9	——
Antimonio	4.4	1167

NOTES — NOTAS

CHAPTER 20/CAPITULO 20
English-Spanish Glossary

ENGLISH-SPANISH GLOSSARY (Alphabetical Listing)

911, Nine-one-one:
911, Nueve uno uno

A few: Pocos

A phase, B phase, C phase:
Fase 1, fase 2, fase 3

Abatement: Anulación
(remoción)

AC current: Corriente CA

Access: Acceso

Access cover:
Tapa de acceso

Accident: Accidente

Account number:
Número de cuenta

Acetylene: Acetileno

Acid: Acido

Acoustical panels:
Paneles acusticos

Acre: Acre

Add: Agregar

Add water: Agregar agua

Additives: Aditivos

Address: Dirección

Adhesive: Pegamento

Adjust: Ajustar

Adjustable wrench:
Llave francesa

After: Despues

Afternoon: Tarde

Again: Otra vez

Aggregate, fines:
Finos de agregado

Aggregates: Agregados

Agree: Convenir

Agreement: Acuerdo

Air compressor:
Compressor de aire

Air conditioning: Aire
acondicionado

Air-entraining agent: Agente
de aeroclusión de cemento

Aisle: Pasillo

Alarm: Alarma

Alert: Alerta, vigilante

Align: Alinear

Alley: Callejón

Allowance: Complemento

Alternate (verb): Alternar

Alternate, alternative:
Suplente

Aluminum: Aluminio

Aluminum siding:
Revestimiento de aluminio

Always: Siempre

ENGLISH-SPANISH GLOSSARY (Alphabetical Listing) *(cont.)*

Ambulance: Ambulancia

Amount: Cantidad, suma

Amperage: El amperaje

Amperes: Amperios

Anchor: Anclar

Anchor bolt: Perno de anclaje

Anchorage point: Punto de anclaje

Affix: Anclar

And: Y

Angle iron: Angulo de hierro

Antibiotic: Antibiotico

Apartment building: Edificio de departamentos

Apply: Aplicar

Apprentice: Aprendiz

Approve: Aprobar

Approved: Aprobado

April: Abril

Arc: Arco

Arc welding: Soldadura al arco

Architect: Arquitecto

Architectural plans: Planos arquitectonicos

Asbestos: Asbesto

Asphalt: Asfalto

Asphalt shingle: Teja de asfalto

Atrium: Atrio

Attach: Conectar

Attic: Ático

August: Agosto

Authorized personnel only: Personal autorizado solamente

Automatic closing device: Dispositivo de cierre automático

Awnings: Toldos

Axe: Hacha

Backflow: Contraflujo

Backhoe: Retroexcavadora

Backing: Soporte

Backup alarm: Alarma de retroceso, Alarma de refuerzo

Backup bar: Varillas adiciónales

Bad: Malo

Bags: Bolsas

Balance: Balance

Ball cock: Valvula de flotador

Ball valve: Llave de flujo

Ballast: Lastre

Bandage: Vendaje

Banding: Banda de ligadura

Bank: Banco

Bank account: Cuenta de banco

Bank name: Nombre de banco

Bar: Barreta

Bar bender: Doblador de varilla

Baricade: Barricada

Base plate: Placa de base

Baseboard: Rodapie

Basement: Sótano

Basin wrench: Llave pico de gansa

Bathroom: Cuarto de baño

Bathroom sink: Lavabo

Bathtub: Bañera

Batter boards: Camilla

Battery: Pila

Bead: Corcon

Beam: Viga

Beam forms: Formas de viga

Bearing wall: Banda de ligadura

Beater: Martillito

Because: Por que

Bedrock: Roca de fondo

Bedroom: Habitación, dormitorio

Before: Antes

Behind: Detras de

Bench mark: Punto de referencia

Better: Mejor

Bid: Oferta

Bit and brace: Taladro de mano

Blade: Hoja

Blankets: Mantas

Blast: Volar

Bleed: Sangar

Bleed (water): Exudarse (agua)

Bleeder valve: Valvula de purga

Bleedwater: Agua de exudación

Blistering: Vesiculación

Block: Bloque, Cubos de construción

Block out: Formar un bloqueo, Bloquear

Blocking: Travesano, Bloqueando

Blower: Sopladora

Blue stakes: Estacas azules

Board: Panel

Boiler: Caldera

Boiler room: Cuarto de calderas

Boiler room: Sitio de la caldera

Bolt: Perno, Metro pernos

Bolt (verb): Enparnar

Bolt patterns: Muestra de pernos

Boltcutters: Cortapernos

Bond: Bono

Bonding agent: Agente de vinculación

Bonding conductor: Cable de anlace

Bonding jumper: Borne de enlace

Boots: Botas

Boss: Jefe

Bottle (oxygen, acetylene): Botella (oxígeno o acetileno)

Bottle caps: Tapas de botella

Boulder: Canto rodado

Boundary: Limite

Bow: Arco

Brace: Jabalcon

Bracing: Apoyo, riostras

Bracket: Brazo

Branch: Ramal

Brass: Bronce

Braze: Soldar en fuerte

Brazing alloy: Aleación para soldar

Brazing flux: Fundente para soldar

Break: Romper, quebrar

Brick: Ladrillo

Bridging: Arriostramiento

Bring: Traer

Broken: Roto, averiado

Broken glass: Cristal roto

Broom: Escoba

Brush: Brocha, cepillar

Bucket: Cubeta, cubo

Build: Construir

Building department: Departamento de construcción

Building inspector: Inspector de obras

Buildup: Acumulación

Bulkhead: Tope de formo

Bull pin: Perno de aliniación

Bullfloat: Plana grande

Buried cable: Cable enterrado

Burn: Quemar

Burn in: Pulir

Burns: Quemaduras

Burr: Rebaba

Bury (verb): Enterrar

But: Pero

Butt joint: Junta de tope

Butt weld: Soldadura a tope

Buy: Comprar

C channel: Perfil en U

C clamp: Grapa en C

Cabinetmaker: Ebanista

ENGLISH-SPANISH GLOSSARY (Alphabetical Listing) *(cont.)*

Cabinets: Gabinetes

Cable tray: Bandeja de portacables

Caissons: Cajones de aire comprimido

Calcium: Calcio

Calculate (verb): Calcular

Call: Llamar

Camber: Comba

Cambered: Combado

Canopy: Toldo

Cantilever: Voladizo

Caps: Tapas

Carpenter: Carpintero

Carpenter's apron: Mandil, delantal

Carpenter's square: Escuadra

Carpet: Alfombra

Carpet layers: Alfombreros

Cart: Carreta

Cartridge fuse: Fusible de cartucho

Cash: Efectivo

Casing: Chambrana

Casing nail: Clavo de cabeza perdida

Cast stone: Piedra moldeada

Cat's paw: Pata de chiva

Caulk: Calafatear

Caulking: Calafeto

Caulking gun: Pistola de calafeto

Caustic: Caustico

Caution: Precaución

Caution tape: Cinta de precaución

Cave-in: Colapso de agujero

Cavity wall: Muro hueco

Ceiling: Cielo

Ceiling joist: Vigueta de cielo

Ceiling suspension: Alambre de la suspensión del techo

Ceiling tiles: Tejas de cielo

Cement: Cemento

Center to center: Centro a centro

Ceramic tile: Azulejos de ceramica

Certificate of occupancy: Certificado de uso

Chain: Cadena

Chain saw: Sierra de cadena

Chain wrench: Llave de cadena

Chalk box: Carrete entizado

Chalk line: Rayero de tiza

Chamfer: Chaflan

Change: Cambio

Change (verb): Cambiar

ENGLISH-SPANISH GLOSSARY (Alphabetical Listing) *(cont.)*

Change order:
Cambio de orden

Channel-lock pliers:
Alicates de extensión

Charge (batteries):
Cargar (las pilas)

Chase: Canaletas

Check: Cheque

Check (verb): Comprobar

Check number:
Número de cheque

Check valve: Valvula
de contraflujo

Chemical: Producto químico

Chicken wire:
Alambre de pollo

Children: Niños

Chimney: Chimenea

Chimney liner:
Revestimiento de chimenea

Chip: Picar

Chisel: Cincel

Choker: Cable ahogador

Chop saw: Serrucho tajadero

Chute: Ducto, canal inclinado

Circuit: Circuito

Circuit breaker: Apagador,
interruptor de circuito

Circuit breaker panel:
Cuadro de cortacircuito

Circular saw:
Serrucho circular, sierra
circular de mano

Circular saw blade: Disco

Cistern: Aljibe

Citizen: Ciudadano

Citizenship: Ciudadanía

Clamp: Grapa, abrazadera

Classroom: Sala de clase

Claw hammer: Martillo chivo

Clay: Arcilla

Clean: Limpio

Clean (verb): Limpiar

Clean cut: Corte limpio

Cleanout: Registro

Cleanout (chimney):
Abertura de limpieza

Clear: Despejar

Clearance: Espacio

Cleat: Tirante de formado

Clinic: Clínica

Clip: Esquilar

Clod: Terron, gelba

Cloth: Tela

Clutch: Cloch, clutch

Coal: Hulla

Coat: Cubrir

Code: Codigo

Cold: Frio

Cold water: Agua fria**

ENGLISH-SPANISH GLOSSARY (Alphabetical Listing) *(cont.)*

Collapse (person): Debilitarse

Collapse (structure): Romperse

Collect: Recoger

Column: Columna

Combination load: Combinación de cargas

Combustible: Combustible

Come (verb): Venir

Come-along: Mordaza tiradora de alambre, trinquete a polea

Communicate: Comunicar

Communication system: Sistema de comunicación

Compact: Compactar

Competent: Competente

Complete: Cumplir

Concrete: Concreto

Concrete broom: Escoba de concreto

Concrete burns: Quemaduras de cemento

Concrete mixing truck: Mezcaldora de concreto sobre camión

Condominium: Condominio residencial

Conductor: Conductor

Conduit: Conducto, canal

Cone ties: Ligaduros cono de pared

Connect: Conectar

Connection: Conexión, unión

Connector: Conector

Conscious: Conciente

Construction adhesive: Pegamento de construcción

Construction schedule: Cronograma de construcción

Contamination: Contaminación

Continuous: Continuo

Contract: Contrato, contratar

Contractor: Contratista

Control joint: Junta de control

Conveyor: Transportador de mecánico

Cool: Enfriar

Coolant: Líquido refrigerante

Cooling tower: Torre de refrigerado

Coordinate: Coordinar

Coping: Albardilla

Copper: Cobre

Cords: Cuerdas

Corner bar: Varilla de esquina

Corners: Esquinas

Cornice: Cornisa

ENGLISH-SPANISH GLOSSARY (Alphabetical Listing) *(cont.)*

Correct, right: Correcto

Corridor: Pasillo

Corrosion resistant: Anticorrosivo

Corrosive: Corrosivo

Coupling: Acoplamiento

Cover: Cubrir

Covering: Recubrimiento

Cracked: Agrietado

Cracks: Rajadas

Crane: Grua

Crane hand signals: Señales de mano de grua

Crane operator: Operador de grua

Crawl space: Espacio angosto

Crescent wrench: Llave de tuercas, llave adjustable

Cripple stud: Mantantes cojos

Cross grain: Fibra transversal

Crown: Vértice

Cubic feet: Pies cubicos

Culvert: Alcantarilla

Curb: Guarnición

Curb and gutter: Arroyo encintado

Curing compound: Compuesto de curado

Curtains: Cortinas

Curve: Curva

Cut: Corte

Cut (verb): Cortar

Cut-off saw: Sierra para cortar

Cut-off valve: Valvula de cierre

Cutting torch: Antorcha

D ring: Anillo en D

Daily report: Reportaje del dia

Damaged: Dañado

Damp: Humedo

Damper: Regulador

Danger: Peligro

Dangerous: Peligroso

Darby: Plana, flatacho

Dark: Oscuro

Date: Fecha

DC current: Corriente CC

Dead end: Terminal, sin salida

Dead load: Carga muerta

Debris: Ruina

December: Diciembre

Deck, decking: Cubierta

Decking: Tablero de lámina

Decompose: Descomponerse

Defective: Defectuoso

ENGLISH-SPANISH GLOSSARY (Alphabetical Listing) *(cont.)*

Degrees (angle): Grados (ángulos)

Dehydrated: Deshidratado

Delay: Retrasar

Demolish: Demoler

Deposit: Deposito

Depth: Profundidad

Destroy: Destruir

Details: Detalles

Determine: Determinar

Dewater: Desaguar

Dial (phone): Marcar (teléfono)

Diameter: Diametro

Die/chaser nut: Dado fracciónario

Diesel: Diesel

Different: Diferente

Difficult: Difícil

Dig: Cavar

Dimensions: Dimensiónes

Directions (North…): Direcciónes (Norte…)

Dirt, earth: Tierra

Dirty: Sucio

Disagree: Discrepar

Discolored: Descolorido

Disconnect switch: Desconectar

Disinfectant: Desinfectante

Distance: Distancia

Distorted: Torcido

Divide: Dividir

Document, record (verb): Documentar

Documentation: Documentación

Dollars: Dolares

Done: Hecho

Door: Puerta

Door bell: Timbre

Door closer: Freno de puerta

Door frame: Marco de puerta

Door jamb: Jamba de puerta

Door sill: Umbral

Door stop: Tope

Doorway: Portal

Dormer: Buharda

Dormitory: Residencias para estudiantes

Double joist: Vigueta doble

Double pole breaker: Interruptor automático bipolar

Dowel: Pasador de varilla

Draft stop: Cierre de tiro

Drag: Halar

Dragline: Excavadora de arrastre

Drain: Desagüe

ENGLISH-SPANISH GLOSSARY (Alphabetical Listing) *(cont.)*

Drain (verb): Drenar

Drain, drainage: Desagüe

Drainage: Drenaje

Draw: Dibujar

Drawer: Cajon

Drawings: Dibujos

Dredge: Draga

Drill: Taladro

Drill (verb): Taladar, agujerear

Drill bit: Broca, mecha

Drilling: Perforación

Drive: Conducir

Drive (a stake): Golpear (una estaca)

Driveway: Vía de acceso

Drunk: Borracho

Dry: Seco, arido

Drywall: Pirca

Duct: Conducto

Duct worker: Trabajador del conducto

Ductility: Ductilidad

Ducts: Conductos

Dump: Basurero, descargar

Dump truck: Volquete, camión de descarga

Dumpster: Tambo de basura

Dust: Quitar el polvo, polvo

Dust control: Eliminación de polvo

Dust mask: Mascarilla para polvo

Dust mop: Trapeador de polvo

Dust pan: Recogedor de polvo

Dusty: Polvoriento

Dwelling: Vivienda

Dwelling unit: Unidad de vivienda

Dynamite: Dinamitar

Ear plugs: Tapones del oído

Early: Temprano

Earth work: Terrapien

Earthquake load: Carga sísmica

East: Este

Easy: Facil

Eave: Alero

Edge forms: Formas de borde

Edger: Orillero

Egress: Salida

Eight: Ocho

Eighteen: Dieciocho

Eighty: Ochenta

Elbow: Codo

Electric drill: Taladro eléctrico

Electrical fixture: Artefactos eléctricos

ENGLISH-SPANISH GLOSSARY (Alphabetical Listing) *(cont.)*

Electrical outlet:
Enchufe, tomacorriente

Electrical plan:
Plano eléctrico

Electrician: Electricista

Electricity: Electricidad

Elevation: Elevación

Elevator: Elevador

Eleven: Once

Embankment: Terraplen

Embed plate: Placa
embutida, placa de anclaje

Emergency: Emergencia

Employer, boss: Jefe

Empty: Vaciar

Enclosure: Cerramiento

End view: Vista de
costado/fondo

Enforce: Hacer cumplir

Engineer: Ingeniero

Enough, sufficient:
Suficiente

Entry: Entrada

Erect: Erigir

Erosion: Erosión

Essential facilities:
Instalaciónes esenciales

Estimate: Estimación

Estimator: Estimador

Evacuate: Evacuar

Even: Uniforme

Examine: Examinar

Exhaust: Escape, extracción

Exhaust fan: Ventilador
de extracción

Exit: Salida

Exit door: Puerta de salida

Expansion bolt: Perno
de expansión

Expansion joint: Junta
de expansión

Explosive: Explosivo

Expose: Exponer

Exposed: Expuesto

Extend: Extender

Extension cord:
Cuerda de extención

Extension ladder:
Escalera de extensión

Exterior: Exterior

Exterior wall: Muro exterior

Extra: Sobra

Eye protection:
Protección de ojos

Fabricate: Fabricar

Fabricated: Fabricado

Façade: Alzado

Face brick: Ladrillo de cara

Face grain: Veta superficial

Face shield: Escudo de cara

Facing brick: Ladrillos para frentes

Factory: Fábrica

Fall: Caerse

Fall harness: Arnes de la caida

Fall protection: Protección de caída

Family: Familia

Fan: Ventilador

Far: Lejos

Fascia: Omposta

Fast: Rápido

Fasten: Atar, sujetar

Fasteners: Anciajes

Faucet: Llave de agua

February: Febrero

Federal taxes: Impuestos federales

Feeder: Alimentador

Feeder cable: Cable de alimentación

Felt: Fieltro

Fence: Cerca

Fertilizer: Fertilizante, abono

Few: Pocos

Fifteen: Quince

Fifty: Cincuenta

File: Lima

Fill: Llenar

Fill, backfill: Relleno

Filler rod: Barra rellenadora

Fillet weld: Soldadura ortogonal

Find (verb): Encontrar

Finish: Acabar

Finished: Acabado

Finishing nails: Clavos sin cabeza

Fire: Fuego

Fire alarm: Alarma de incendio

Fire Code: Codigo de Incendios

Fire extinguisher: Extintor

Fire proofing: Ignifugación

Fire sprinklers: Regaderas de fuego

Fire stop: Tope antifuego

Firebrick: Ladrillo de fuego

Fireplace: Chimenea

First: Primera

Fish tape: Cinta pescadora

Fitting: Accesorio

Five: Cinco

Fix (verb): Arregle

Fixture: Artefacto

Flag stake: Estaca de bandera

Flagger: Persona que da señales

Flamable: Inflamable

Flammable liquid: Liquido inflammable

Flash burns: Quemadura de relampago

Flashing: Cubrejuntas, tapajuntas

Flashlight: Linterna

Flat: Plano

Flat bed truck: Camión de tarima

Flat roof: Techo plano

Flat ties: Ligadura plana

Flat weld: Soldadura de plano

Flat work: Pieza plana

Flexible conduit: Conducto portacables flexible

Float: Plana

Floodlight: Iluminación

Floor: Piso

Floor deck: Plataforma

Floor girder: Viga principal

Floor joist: Vigueta

Floor tiles: Baldosas

Flooring: Revestimientos para pesos

Flue: Conductos de humo

Fluorescent: Fluorescente

Flush: A nivel

Flush weld: Soldadura a pano

Flux: Fundente

Fly ash: Cenizas volantes

Fold: Doblar

Folding partition: Tabique plegable

Footing: Zapata, zapata de cimentación

Footing form: Forma de zapata de cimentación

For: Por, para

Foreman: Capataz

Form: Forma

Form oil: Aceite de formas

Form ply: Madera laminada para formar

Forty: Cuarenta

Foundation: Fundación

Foundation wall: Muro de fundación

Four: Cuatro

Fourteen: Catorce

Fragile: Frágil

Frame: Marco, estructura, sujetar, formar

Framework: Armazón

Framing: Intramado

Framing square: Escuadra

Freezing: Congelante

Frequency: Frequencia

ENGLISH-SPANISH GLOSSARY (Alphabetical Listing) *(cont.)*

Fresh/green: Fresco/verde

Friday: Viernes

Friend: Amigo

From: De

Front: Frente, delante

Front view: Vista de frente

Frost: Helada

Fumes: Gases

Funnel: Embudo

Furnace: Horno

Furring, furred: Enrasado

Fuse: Fusible

Fuse box: Caja de fusibles

Gable: Hastial

Gable roof: Techo a dos agues

Galvanized: Galvanizado

Gap: Brecha

Garbage: Basura

Garden hose: Manguera de jardín

Gas main: Conducto principal de gas

Gasoline: Gasolina

Gate: Puerta de cerco

Gauge (instrument): Monometro

Gauge (thickness): Caliber

Gauges: Indicadores, Medidores, Metros

Generator: Generador

Geotechnical engineer: Ingeniero de geotécnico

Girder: Viga principal

Give (verb): Dar

Glass: Vidrio

Glazing: Vidriado

Gloves: Guantes

Glue: Pegamento

Glue, adhesive: Adhesiva

Go (verb): Ir

Go (you): Ve

Goggles: Anteojos

Good: Bien

Goodbye: Adiós

Grade: Grado, Nivelar

Grade beam: Viga de fundación

Grade stick: Palo de grado

Graded lumber: Madera elaborada

Grader: Máquina niveladora

Gravel: Grava

Grease: Engrasar

Grease interceptor: Interceptor de grasas

Grease trap: Collector de grasas

Greasy: Grasiento

ENGLISH-SPANISH GLOSSARY (Alphabetical Listing) *(cont.)*

Green card: Tarjeta de residencia

Grid lines: Lineas de rejilla

Grille: Rejilla

Grind: Moler

Grind (metal): Afilar

Grinder: Afilador, amoladora, moledora

Grinding wheel: Disco de amoladora

Grip: Agarre

Grommet: Arandela

Groover: Ranuradora

Gross area: Area total

Ground: Conectar a tierra

Ground bar: Bandeja a tierra

Ground connection: Conexión de tierra

Ground elevation: Rasante

Ground fault circuit Interruptor (GFCI): Interruptor fusible de seguridad a tierra

Ground lead: Cable conductor de tierra

Ground rod: Barra a tierra, vara a tierra

Ground wire: Cable a tierra

Grout: Lechada

Guard rail: Barandal

Gusset: Cartabon

Gutter: Gotera

Gymnasium: Gimnasio

Gypsum: Yeso

Gypsum board: Panel de yeso

Hack saw: Sierra de arco para metal

Hallway: Vestíbulo

Hammer: Martillo

Hammer drill: Rotamartillo

Hand (door): Colgar (puerta), mano

Hand brush: Cepillo de mano

Hand grinder: Afilador de mano

Hand jointer: Marcador de juntas de mano

Hand saw: Serucho de mano, sierra de mano

Hand signal: Señal de mano

Hand stone: Piedra de mano

Hand tools: Herramientas de mano

Handle: Manija

Handle (verb): Manipular

Handrails: Barandales

Hang: Colgar

Hangers: Ganchos

Hard: Duro, firme

Hard drawn copper: Cobre estirado en frio

ENGLISH-SPANISH GLOSSARY (Alphabetical Listing) *(cont.)*

Hard hat: Casco

Hard hat area: Area de cascos

Hat channel: Perfil a sombrero

Hatch: Compuerta

Haul: Transportar, acarrear

Have (verb): Tener

Hazard: Peligro

Hazardous: Peligroso

Hazardous communications: Comunicaciónes peligrosas

He: El

Head protection: Protección de cabeza

Header: Cabezal

Health insurance: Seguro de salud

Healthy: Sano

Heat: Calor

Heat stroke: Insolación

Heater: Calefactor, estufa

Heating: Calefacción

Heavy: Pesado

Height: Altura

Hello: Hola

Help: Ayudar

Helper: Ayudante

Here: Aquí

Hickey bar: Dobladora portátil

High spot: Punto alto

High-rise building: Edificio de gran altura

Hinges: Bisagras

Hip: Lima

Hip roof: Techo de cuarto agues

Hire, employ (verb): Contratar, emplear

Hit: Pegar

Hoe: Azadón

Hold: Detener

Hole: Agujero

Holiday: Día feriado

Honeycombed: Panaleado

Hood (kitchen): Campana (cocina)

Hook: Gancho

Hose: Manguera

Hose bibb: Grifo de manguera

Hospital: Hospital

Hot: Caliente

Hot bus bar: Bandeja de carga

Hot water: Agua caliente

Hour: Hora

House: Casa

How: Como

ENGLISH-SPANISH GLOSSARY (Alphabetical Listing) *(cont.)*

Hub: Eje

Hub valve: Valvula de cubo

Hurt: Herido

HVAC: Calefacción, ventilación y aire acondiciónado

Hydrant: Boca de riego

Hydration: Hidratación

I: Yo

I beam: Viga doble T

I'm sorry: Lo siento

Ice: Hielo

Illegal: Ilegal

Impact wrench: Arranque de impacto

Impedance: Impedencia

In: En

Incline: Declive, inclinación

Incompetent: Incompetente

Incorrect, wrong: Incorrecto

Injury: Lesión

Insert: Insertar

Inside: Adentro

Inspect: Examinar

Inspection: Inspección

Inspector: Inspector

Install: Instalar

Insulation: Aislamiento/ la insulación

Insurance: Seguro

Insurance: Aseguranza

Interior: Interior

Interior finish: Pirca

Interlayment: Capa intermedia

Interlocking: Enclavamiento

Interpreter: Interprete

Jack: Gato

Jackhammer: Martillo neumático

Jamb: Jamba

Jammed: Atorado

January: Enero

Jigsaw: Sierra de vaiven

Job: Trabajo

Job box: Caja de trabajo

Job report: Reporte de trabajo

Job site: Lugar de la obra

Joint: Junta

Joint compound: Pasta de muro

Jointer: Cepillo automático

Jointer plane: Cepillo de mano

Joint-filler trowel: Paleta de relleno

Joist: Vigueta

Joist bridging: Arriostramiento de vigueta

ENGLISH-SPANISH GLOSSARY (Alphabetical Listing) *(cont.)*

Joist hanger: Soporte de vigueta
Journeyman: Oficial
July: Julio
Junction box: Caja de conexiónes de empalme
June: Junio
Junk: Junce, chatarra
Keel: Creyon de Madera
Keep: Guardar
Keep out: Mantengase fuera
Kettle: Caldera
Key: Llave
Key valve: Valvula de llave
Keystone: Clave
Keyway: Llave de cimentación
Kicker: Puntal
King post: Poste principal
Kiosk: Quiosco
Kitchen: Cocina
Knee pads: Rodilleras
Knife: Cuchilo
Knock over: Tumbar
Knockout: Agujero ciego
Knot: Nudo de Madera
Know (verb): Saber
Kraft paper: Papel Kraft
Label: Etiquetar
Ladder: Escalera

Lamp: Lampara
Land fill: Deposito de basura
Landing (stair): Descanso de escaleras
Landscaper: Paisajista
Lanyard: Acollador
Lap siding: Revestimiento de tablas con traslape
Lap weld: Soldadura a solape
Large: Grande
Laser: Laser
Last: Ultima
Latch: Aldaba
Late: Atrasado, tarde
Later: Despues
Lateral (pipe): Ramal lateral
Lath: Liston
Lawnmower: Cortado de césped, cortacesped
Lay (carpet): Poner (alfombra)
Lay out: Croquis, hacer disposiciónes
Leader (pipe): Tubo de bajada
Leak: Fugs, gotera
Leather gloves: Guantes de cuero
Legal: Legal
Length: Longitud

Less: Menos

Lethal: Mortal

Level: Nivel, anivelado, palanca

Lien: Derecho de retención

Lift: Elevar

Lift (verb): Levantar

Light: Liviano

Light bulb: Foco

Lighter: Encendedor

Lights: Luces

Limbs: Miembros

Lime putty: Mastique de cal

Limestone: Caliza

Lining: Recubrimiento

Link, linkage: Enlace, tirante

Lintel: Dintel

Little: Poquito

Live load: Cargas vivas

Live wires: Alambres vivos

Load-bearing joist: Viga de carga

Loaded area: Area cargada

Loader: Cargador

Loan: Prestamo

Local taxes: Impuestos locales

Locate: Localizar

Located: Localizado

Lock: Cerradura, cerrar (bajo llave)

Lock (verb): Cerrar

Lockers: Gavetas

Lockout/tagout: Cierre

Longitudinal and transverse bars: Varillas longitudinales y transversales

Look (verb): Mirar

Loop: Lazadas

Loose: Flojo

Lot: Terreno, lote

Loud: Alto, Ruidoso

Louver: Celosia

Low spot: Punto bajo

Lower: Bajar

Lubricate: Lubricar

Lumber: Madera de construcción

Machine: Máquina

Machinist: Maquinista

Magnesium: Magnesio

Main: Principal, matriz

Main breaker: Interruptor automático principal

Main power cable: Cable principal

Main vent: Respiradero matriz

Maintenance: Mantenimiento

Mall: Centro commercial

Mallet: Mazo

Man: Hombre

Manager: Gerente

Manhole: Pozo de confluencia, boca de inspección, boca de acceso, pozo de entrada

Mansard roof: Mansardara

Manual pull station: Alarma de incendio manual

Manufacture: Fabricar

Manufacturer: Fabricante

Many: Muchos

March: Marzo

Margin trowel: Cuchara para cemento

Mark: Marcar

Mask: Mascara, careta

Masking tape: Cinta de enmascarar

Mason: Albañil

Mason's trowel: Paleta de albañil

Masonry: Albañileria, mamposteria

Mastic: Mastique

Mat: Estera

Matches: Fosforos

May: Mayo

Means of egress: Medios de salida

Measure: Medir

Mechanic: Mecánico

Mechanical plan: Plano mecánico

Medicine cabinet: Botiquín

Medicare: Medicare

Medicine: Medicación, medicina

Metal deck: Plataforma metálica

Metal gutter: Gotera

Metal pile: Pila de metal

Metal scribe: Trazador de metal

Meter: Medidor

Meter (measuring device): Metro, medidor

Mezzanine: Entresuelo

Middle: Medio

Minute: Minuto

Missing: Le falta

Mitre box: Caja de corte a ángulos

Mitre saw: Sierra de retroceso para ingletes

Mix: Mezclar

Modification: Modificación

Modify: Modificar

ENGLISH-SPANISH GLOSSARY (Alphabetical Listing) *(cont.)*

Moist curing: Curado con humedad

Moisture: Humedad

Molding: Moldura

Monday: Lunes

Money: Dinero

Mop: Trapeador

Mop (verb): Trapear

Mop bucket: Cubeta del trapeador

More: Mas

Morning: La mañana

Mortar: Molcajete, mortero

Mortise: Ranura

Motivated: Motivado

Motor: Motor

Move: Mover

Move (verb): Mueva

Moving partition: Tabique movible

Moving walkway: Caminos moviles

Much: Mucho

Mud: Lodo

Muddy: Fangoso (tierra)

Mullion (door): Larguero central

My: Mi

Nail: Clavar

Nail gun: Clavador neumático

Nail hammer: Martillo de clavos

Nail set: Botador de clavos

Nails: Clavos

Name: Nombre

Natural gas: Gas natural

Near: Cerca, próximo

Need (verb): Necesitar

Neoprene float: Flota de neoprene

Neutral: Neutro

Neutral bar: Bandeja neutral

Neutral conductor: Cable neutro

Never: Nunca

Nicked: Muescado

Night, evening: Noche

Nine: Nueve

Nineteen: Diecinueve

Ninety: Noventa

No: No

No smoking area: Zona de no fumar

North: Norte

Notes: Notas

Notice: Aviso

Notice to proceed: Aviso de proceder

November: Noviembre

Now: Ahora

Nut: Tuerca

Nuts: Tuercas

Oak: Roble

Occupancy: Destino, tenencia

Occupant load: Número de ocupantes

October: Octubre

Of: De

Off: Fuera de

Office: Oficina

Offset: Desplazamiento, desvio

OK: OK

Old: Viejo

On: En, sobre

On grade: De altura

On line: En línea

On time: A tiempo

Once: Una vez

One: Uno

One hundred: Cien

One million: Un millón

One quarter: Un cuarto

One thousand: Mil

Only: Solamente

Open: Abra, abrir

Open (verb): Abrir

Open air: Aire libre

Opening: Abertura

Or: O

Order: Ordenar

Ornamental metal: Metal ornamental

Ornamental railing: Pasamanos ornamental

Outlet box: Caja de enchufe, Caja de tomacorriente

Outside: Afuera

Over: Encima

Overhang: Voladizo, vuelo, alero

Overhead view: Vista de arriba

Overhead weld: Soldadura de encima

Overhead work: Trabajo de arriba

Overlap: Traslapo

Overlap (verb): Traslapar

Owe: Deber

Owner: Dueño

Oxygen: Oxígeno

Pain: Dolor

Paint: Pintar

Paint brush: Brocha de pintura

Paint mark: Pintura de marcar

Paint roller: Rodillo de pintura

Paint thinners: Diluyentes de pintura

Painter: Pintor

Panel: Panel

Paneling: Empanelado

Paper dispensers: Dispensadores de papel

Parallel: Paralelo

Parapet wall: Pared de parapeto

Parking lot: Estaciónamiento

Particle board: Tablero prensado

Partition: Partición

Passport: Pasaporte

Patch: Parchar

Payday: Día de paga

Payee: Beneficiario

Payment: Pago

Pea gravel: Gravita

Pedestrian walkway: Calzada

Pen: Pluma

Pencil: Lápiz

Penetration: Penetración

Pension: Pensión

Penthouse: Sobradillo

Performance: Desempeño

Performance bond: Bono de ejecución

Perimeter guard: Guardado de proximidad

Permit: Permiso

Perpendicular: Perpendicular

Phase: Fase

Philips screwdriver: Desarmador cruz

Phone number: Número de teléfono

Pick axe: Zapapico

Pick up: Recoger

Pick-up truck: Camioneta

Pier: Estribo

Pier and column forms: Formas de estribo y columna

Pilaster: Pilistra

Pile: Pila

Pile (verb): Apilar

Pinch point: Punto de pellizco

Pine: Pino

Pins: Pernos

Pipe: Pipa

Pipe wrench: Llave de tubo

Pipe, piping: Cañeria, caño, tubo

Pipeline: Tubería

Pipes: Pipas

ENGLISH-SPANISH GLOSSARY (Alphabetical Listing) *(cont.)*

Pit: Hoyo, pozo

Pitch: Pendiente del techo

Place: Colocar

Plan: Planear

Plane: Cepillo

Planking: Entablonado

Planks: Tablones

Plant (verb): Plantar

Planter: Plantador

Plants: Plantas

Plaster: Azotado, jaharro

Plastering: Revoque, enclucido

Plastering trowel: Llana de emplastar

Plastic: Plástico

Plastic insulator: Aislante plástico

Plasticizer: Plasticizador

Plate: Placa

Plate girder: Viga de alma llena

Plenum: Pleno, camara de distribución de aire

Pliers: Alicates, pinzas

Plot: Trazar

Plug fuse: Fusible de rosca

Plug in: Enchufar

Plumb: Plomada

Plumb (verb): Plomear

Plumb bob: Plomo

Plumber: Fontanero, plomero

Plumbing: Instalación de hidráulicas y sanitarias, plomería

Plumbing appliance: Mueble sanitario

Plumbing plan: Plano de plomeria

Plunger: Destapacaños, sopapa

Ply: Capa

Plywood: Chapeado

Point (shape): Punta

Point (verb): Señalar

Poison: Veneno

Pole: Poste

Polish: Pulir

Portable: Portátil

Portable partition: Tabique portátil

Portland cement: Cemento Portland

Position: Posiciónar

Positioning chain: Cadena de disposición

Post: Poste

Pound: Apilar

Pour (verb): Echar

Pour coat: Capa de colada

Powder nailer: Pistola de cartuchos para fijación

Power: Potencia

Power doors: Puertas mecanicas

Power lines: Lineas de energía

Power strip: Zapatilla electrica

Power supply: Fuente de alimentación

Power trowel: Llana mecánica

Precisely (time): En punto (tiempo)

Preheat: Precalentar

Premises: Local, sitio

Preparation: Preparación

Press: Prensar, presionar

Pressure: Presión

Prestressed concrete: Hormigon preesforzado, hormigonprecargado

Price: Precio

Primed: Imprimado

Primer: Imprimidor

Private: Privado

Problem: Problema

Promise: Promesa

Propane: Propano

Property: Propiedad

Property boundary line: Línea de límite de la propiedad

Property line: Línea de propiedad

Proportion: Proporción, dimensiónar

Proportioned: Dimensiónado

Protection: Protección

Protruding: Saliente

Protruding nails: Clavos salientes

Provision: Disposición

Pry bar: Barra de palanca

Public safety: Seguridad pública, protección al público

Public utilities: Sevicios publicos

Puddle: Charco

Puddle weld: Soldadura a charco

Pull: Jalar

Pump: Bomba

Pump (verb): Bombear

Pump truck: Bombero de concreto

Punches: Punzones

Push: Empujar

Push broom: Escobon

Put: Colocar

Put (verb): Poner, colocar

ENGLISH-SPANISH GLOSSARY (Alphabetical Listing) *(cont.)*

Put on: Poner

Putty coat: Enlucido

Putty knife: Espatula de masilla

Queen post: Columna

Quicklime: Cal viva

Quickly: Rapido, rapidamente

Quickly: Pronto

Quicksand: Arena movediza

Quiet: Tranquilo, silencio

Quota: Cuota

Rabbet: Muesca, ranura

Raceway: Conducto eléctrico

Rack: Cremaliera, tarima, gualdrapear

Radial arm saw: Serrucho guillotina

Radius: Radio

Rafter: Cabio

Rags: Trapos

Rail: Cremallera, barandal

Railing: Barandal, barra, carril

Rainy: lluvioso

Raise: Levantar, alzar

Rake: Rastrillo, rastillar

Ram: Impelar

Ramp: Rampa

Range outlet: Tomacorriente para estufa, enchufe para estufa

Rate: Relación, razón

Rating: Clasificación

Ratio: Relación, cociente

Read (verb): Leer

Ready: Listo

Rebar: Barra de refuerzo, varilla

Rebar bender: Doblador de varilla

Rebar caps: Tapadera de barilla, tapas de varilla

Rebar shears: Tijeras de varilla

Receptacle, plug: Enchufe, clavija

Reciprocating saw: Sierra alternativa

Record: Registrar

Recover: Recuperar

Recycle: Reciclar

Red hot: Fosforescente, candente

Redwood: Madera de secoya

Reflectors: Reflectores

Reformatory: Reformatario

Region: Región, tramo

Reglet: Regleta

Regulator: Regulador

Regulator valves: Valvulas de regulación

Reinforce: Reforzar

Reinforced masonry: Mamposteria reforzada

Reinforcement: Refuerzo

Reinforcing steel: Acero para reforzar

Rejected: Rechazado

Release: Descarga, liberación

Release (verb): Soltar

Relief valve: Valvula de alivio, llave de alivio

Remove: Quitar

rent (verb): Rentar

Rental: Alquiler, alquilar

Repair: Reparación

Repair (verb): Reparar

Repeat (verb): Repitir

Replace: Reemplazar, reponer

Report: Reportaje

Require: Requerir

Rescue: Salvar

Resistance: Resistencia

Respirator: Respirador

Responsibility: Responsabilidad

Responsible: Responsable

Restraint : Sujetador

Retardant: Retardador

Retractable reel: Carrete retractable

Retrofit: Retroajuste

Reverse threaded: Reverso roscado

Revolving door: Puerta giratoria

Ridge: Cresta

Ridge board: Tabla de cumbrera

Ridge tile: Tejas para cumbrera

Rig: Aparejar

Right: Derecho

Right angle: Angulo recto

Right angle drill: Taladro de angulo recto

Rim beam: Viga del borde

Rinse: Enjuagar

Ripper: Destripador

Riser (pipe): Tubo vertical

Riser (stair): Contrahuella

Rivet: Remanche

Road: Camino, carretera

Road base: Base de pavimento

Rock: Roca, piedra

Roll: Rodar

Roller: Rodillo

Roof: Techo, azotea

Roof covering: Revestimiento de techo, cubierta de techo

Roof deck: Cubierta de techo

Roof drain: Desagüe de techo

Roof sheathing: Entablado de techo, entarimado de tejado

Roof tile: Teja

Roofer: Techero

Roofing felt: Tela asfaltica, felpa

Roofing square: Cuadro de cubierta de techo

Room: Cuarto, sala

Rough: Aspero

Rough finished: Acabado ordinario

Rough opening: Abertura bosqueja

Rough sill: Umbral bosquejo

Rough-in: Instalación en obra negra, instalación de obra gruesa

Router: Fresadora, contorneador

Rub: Frotar

Rubber boots: Botas de goma

Rubber gloves: Guantes de goma

Rubbish: Desperdicios

Rubbish chute: Ducto de basura

Rubble: Escombro

Rules: Reglas

Run (use): Correr (usar)

Runners: Largueros

Runoff: Agua de desagüe

Rust: Moho

Safe: Ileso, seguro

Safely: Con seguridad

Safety glazing: Vidriado de seguridad

Safety policy: Póliza de seguridad

Salary: Salario

Salvage: Salvar

Same: Uniforme, mismo, igual

Sand: Arena

Sand (verb): Lijar

Sand paper: Lija

Sander: Lijadora

Sandstone: Arenisca

Sanitary: Sanitario

Sanitary sewer: Sistema de alcantarilla sanitario

Sanitation: Higiene, sanidad

Saturday: Sabado

Saw: Serrucho

Saw (verb): Aserrar

Saw guard: Protector de serrucho eléctrico

Saw horse: Banqueta de aserrado

Scaffold: Andamio

Scaffolding: Andamiaje, andamios

Schedule: Horario

Schedules: Horarios

Scissor lift: Plataforma hidráulica

Scoop: Recoger

Scope: Alcance

Scrap: Desechar

Scrape: Raspar

Screed rod: Formas para pieza plana

Screeds: Botas de goma

Screw: Tornillo, atornillar

Screw connector: Conector con tornillo

Screw gun: Pistola desarmadora

Screwdriver: Destornillador, desarmador

Screws: Tornillos

Scrub: Fregar

Seal: Sellar

Sealant: Sellador

Seasoned wood: Madera de estaciónada

Secretary: Secretario, secretaria

Section: Sección

Secure: Asegurar

Secured: Asegurado

Securely: Seguramente

Security system: Sistema de seguridad

Seed: Sembrar, semilla

Self-closing: Autocierre

Self-drilling screws: Tornillos autoperforantes

Self-ignition: Auto-ignición

Self-luminous: Autoluminoso

Self-tapping screws: Tornillos autorroscantes

Sell: Vender

September: Septiembre

Septic tank: Fosa septica

Series: Series

Service entrance neutral: Cable principal neutro

Set: Puesto

Set up: Fraguarse

Setback: Retiro

Setting time: Tiempo de fraguado

Seven: Siete

ENGLISH-SPANISH GLOSSARY (Alphabetical Listing) *(cont.)*

Seventeen: Diecisiete

Seventy: Setenta

Sewage: Aguas negras

Sewer: Alcantarilla

Shackle: Grillete

Shaft: Recinto

Sharp: Filoso

Sharp edge: Borde agudo

She: Ella

Sheathing: Entablado

Sheer wall: Muro cortante

Sheet: Pliego, chapa, plancha

Sheet copper: Lamina de cobre

Sheet metal: Lamina metálica, chapa metálica

Sheet metal shears: Tijeras para metal

Sheeting: Revestimiento, laminado

Sheetrockers: Yeseros

Shelf: Repisa

Shell: Cascara, cubierta

Shim: Acunar

Shine: Brillar

Shingle: Teja

Shiny: Brilliante

Shock: Sacudida de eléctrica

Shoot (a line): Tirar (una linea)

Shop: Taller

Shop drawings: Dibujos de taller

Shored construction: Construcción apuntalada

Shoring: Escoramiento

Shoring: Puntales

Shovel: Pala

Shovel (verb): Traspalar

Show window: Vitrina

Showcase: Armario de exhibición

Shower stall: Ducha, regadera

Showerhead: Regadera

Shrubs: Arbustos

Shut off: Apagar

Shutoff valve: Valvulas de cierre

Sick: Enfermo

Sidewalk: Acera

Siding: Forrado, revestimiento

Signal: Señalar con la mano

Signature: Firma

Silica sand: Arena de silicona

Sill: Umbral

Sill cock: Grifo de manguera

Sill plate: Solera inferior

Single pole breaker: Interruptor automático unipolar

Sink: Fregadero, hundirse

Site plans: Planos se sitio

Six: Seis

Sixteen: Dieciseis

Sixty: Sesenta

Skylight: Lucernario

Skylight: Tragaluz, claraboya

Slab: Losa

Slab forms: Formas de losa

Slag: Escoria

Slag hammer: Martillo de escoria

Slate shingle: Teja de pizarro

Sledgehammer: Marro, mazo

Sleeve: Camisa, manga

Sleever bar: Palanca de cola

Slick: Liso

Sling: Eslinga

Slip: Deslizar

Slippery: Resbaladizo

Sliver: Astilla

Slope: Cuesta, inclinación, pendiente, talud

Slow: Despacio

Slowly: Lento, lentamente

Slump: Sección, asentamiento

Slushy: Fangoso (nieve)

Small: Pequeño

Smash: Allanar

Smoke: Humo

Smoke barrier: Barrera antihumo

Smoke detector: Sensor de humo

Smoke-tight: Impermeables al humo

Smooth: Liso

Snake: Serpiente

Snap tie: Atadura instantanea

Snow: Nieve

Soap: Jabón

Soap stone: Jaboncillo

Social security number: Número de seguro social

Sod: Cesped

Soffit: Sofito

Soft: Blando, suave

Soil: Suelo

Soil engineer: Ingeniero de tierras

Soil pipe: Tubo bajante de aguas negras

Soil stack: Bajante sanitaria

Soil type: Tipo de suelo

ENGLISH-SPANISH GLOSSARY (Alphabetical Listing) *(cont.)*

Soldering torch: Soplete

Sole plate: Placa de base

Soleplate: Solera

Solid: Solido

South: Sur

Spackle: Junta de cemento

Span: Abarbetado

Spandrel: Jacena exterior, timpano

Sparks: Chispas

Speak (verb): Hablar

Spigot: Llave, grifo, canilla

Spike: Clavo especial para madera

Spiked: Clavado

Spiral stairs: Escaleras de caracol

Splice: Empalme, translape, junta

Splice (verb): Empalmar

Splint: Tablilla

Sponge: Esponja

Sponge float: Flota de espona

Spot footing: Zapata de columna

Spray: Rociar

Spray bottle: Rociador

Spray paint: Pintura rociada

Sprayer: Rociador

Spreader: Seperador

Spring: Resorte

Sprinkle: Asentamiento

Sprinkler: Rociador

Sprinkler control box: Caja de controles de regadera

Sprinkler head: Rociador

Sprinkler system: Sistema de rociadores

Sprinklers: Regaderas

Spud wrench: Llave de cola

Square: Escuadra, cuadrado

Square trowel: Llana

Squeegee: Enjuagador de vidrio

Stability: Estabilidad

Stabilize: Estabilizar

Stack: Tuberia vertical bajante

Stack (verb): Apilar

Stack vent: Respiradero de bajante

Stainless steel: Acero inoxidable

Stair forms: Formas de escalera

Stair stringer: Zanca de escalera

Stairs: Escalones

Stairwell: Recinto de escaleras

Stand (verb): Parar, erguir

Standpipe: Columna hidrante

Standpipe system: Sistema de columna hidrante

Stapler: Engrapadora

Stapling gun: Engrapadora automática

Start (verb): Empezar

State taxes: Impuestos estatales

Stay off: Quedarse fuera

Steam: Vapor

Steel: Acero

Steel studs: Montantes, pies derechos de acer

Steel toe boots: Botas prectores de la punta del pie

Step ladder: Escalera de tijera

Steps: Gradas

Sterile: Estéril

Stick: Pegar

Stirrup: Brida

Stone: Piedra

Stop (verb): Parar

Storage room: Cuarto de almacenamiento

Store: Tienda

Store (verb): Almacenar

Storm drain: Drenaje para tormentas

Straight: Derecho

Strap: Fleje

Strap wrench: Llave de correa, llave de cincho

Strapping: Flejes

Street: Calle

Striker: Percutor

String line: Línea de hilo

Strong: Fuerte

Structural plans: Planos estructurales

Structural steel: Acero structural

Structure: Estructura

Strut: Puntal

Stucco: Estuco

Stuck: Pegado, travado

Stud finder: Buscador be montantes

Stud track: Carril de montante

Stud wall: Muro con montantes

Studs: Montantes, pies derechos de madera

Subcontractor: Subcontratista

Subfloor: Subsuelo

Subgrade: Subsuelo

Submit: Someter

Sump: Sumidero

Sump pump: Bomba de sumidero

Sump vent: Respiradero de sumidero

Sunday: Domingo

Superintendent: Superintendente

Supervisor: Supervisor, supervisora

Surface water: Agua superficial

Survey: Topografar

Survey (verb): Deslindar

Surveyor: Topografo

Suspended ceiling: Falso plafon, cieloraso suspendido

Sweep: Barrer

Swelling: Expansión

Swimming pool: Piscina (de natación), alberca

Swing: Oscilación

Swing (verb): Oscilar

Swinging door: Puerta pivotante

Switch: Interruptor, apagador

Switch plate: Placa de interruptor

Symbols: Simbolos

Table saw: Serrucho de mesa

Tag: Etiqueta

Tag line: Línea de guía

Take (verb): Tomar

Take off: Extracción de los planos

Tamp: Apisonar

Tamper: Pisón

Tandem: Tandem

Tanks: Tanques

Tape measure: Cinta de medir

Taping compound: Pasta de muro

Tar: Alquitran, brea, chapopote

Tar paper: Papel de brea

Tarp: Lona

Taxes: Impuestos

Teamwork: Trabajo de cooperación

Tear apart: Desarmar

Technician: Técnico

Tee: T, injerto

Telephone: Teléfono

Television system: Sistema de televisión

Templates: Plantillas

Temporary: Provisiónal

Temporary power: Energía temporaria

Ten: Diez

Tenant: Inquilino

Tension: Tensión

Tents: Carpas

Termite protection: Proteción contra termitas

Test: Probar

Tetanus shot: Inyección contra el tétanos

Texture: Textura

That: Ese, esa

Thawing: Descongelación

Them: Ellos

Theodolite: Teodolito

There: Alla

Thermostat: Termostato

Thin: Delgado

Thirteen: Trece

Thirty: Treinta

This: Este

Thread: Hilo

Three: Tres

Threshold: Umbral

Throw away: Descartar

Thursday: Jueves

Tie: Amarra, ligadura

Tie (verb): Atar

Tie off: Atarse seguramente

Tie wire: Alambre de atadura

Ties: Atadura

Tight: Apretado, aprieta

Tighten: Apretar

Tile setter: Azulejero

Tin: Lata, chapa,

To: A

Today: Hoy

Toe board: Tabla de pie

Toe nail: Clavo oblicuo

Toilet: Inodoro, sanitario

Tomorrow: Mañana

Tongue and groove: Machihembrado

Tool handles (poles): Mangos de herramientas (tubos)

Tool lockup: Cuarto de herramientas llavado

Tools: Herramientas

Top plate: Placa superior, placa

Topographic map: Mapa topográfico

Topsoil: Tierra vegetal

Torch: Antorcha

Torque: Apretar (una tuerca)

Torque wrench: Llave dinamometrica

Total station: Transito de computadora

Tracer: Alambre testigo

Track hoe: Encarrilada

Traction: Tracción

ENGLISH-SPANISH GLOSSARY (Alphabetical Listing) (cont.)

Traffic awareness: Conocimiento de tráfico

Train, teach: Entrenar

Training: Entrenamiento

Transformer: Transformador

Transit: Transito

Transplant: Transplantar

Trap: Sifon

Trap seal: Sello de trampa hidráulica

Trash: Basura

Treated lumber: Madera de construcción tratada

Tree: Arbol

Trees: Arboles

Trench/ditch: Zanja

Trim: Molduras

Tripod: Tripode

Truck: Camión

Truck bed: Caja de camión

Truck tailgate: Puerta trasera de camión

Truss: Armadura de cubierto

Tuesday: Martes

Tunnel: Túnel, cavar un túnel

Turn buckle: Tensor

Turn off: Apagar

Turn on: Prender

Tweezers: Pinzas pequeñas

Twelve: Doce

Twenty: Veinte

Two: Dos

Two by four: Dos por cuatro

Two by twelve: Dos por doce

Unbalanced fill: Relleno sin consolidar

Unbalanced loads: Cargas no balanceadas

Under: Debajo

Undercut: Resquicio

Underground: Subterráneo

Underground lines: Líneas subterráneas

Underlayment: Suelo subyacente

Uneven: Desigual

Union: Unión, sindicato

Unit price: Precio de módulo

Unload: Descargar

Unsafe building: Edificación insegura

Unshored construction: Construcción no apuntalada

Unstable ground: Terreno inestable

Uplift (wind): Remonte

Urinal: Urinario

Us: Nosotros

Utility knife: Navaja de utilidad

Utility lines: Líneas de servicios publicos

V joint: Junta en V

Vacuum: Vacío, aspiradora

Vacuum (verb): Pasar la aspiradora

Vacuum breaker: Interruptor de vacío

Valuation: Valuación

Value: Valor

Valve: Valvula

Valve-seat wrench: Llave de asiento de valvula

Veins: Venas

Veneer: Revestimiento

Vent (verb): Ventilar, evacuar

Vent shaft: Recinto de ventilación

Vent stack: Respiradero vertical

Vent system: Sistema de ventilación

Ventilate: Ventilar

Venting system: Sistema de evacuación

Verify: Aseguarse, Verificar

Vertical pipe: Tubo vertical

Vertical weld: Soldadura vertical

Vestibule: Vestíbulo

Vests: Chalecos

Vibrate: Vibrar

Vibrator: Recibo del concreto

Vice bench: Torno de banco

Vinyl siding: Revestimiento de vinilo

Visa: Visa

Vise: Morsa, tornillo de banco

Vise-grip pliers: Alicates de presión, pinzas perras

Visqueen: Pelicula de polieteno

Void space: Espacio vacío

Voltage: Voltaje

Volts: Voltios

Wainscoting: Friso, alfarje

Wait: Esperar

Waler: Larguero

Waler loops: Ligaduras de larguero

Walk: Caminar

Walkie talkie: Transceptor de portátil

Walk-in cooler: Frigorífico

Walkway: Acera

Wall: Pared

Wall brace: Apoyo de pared

Wall line: Línea de la pared

Wall sheathing: Entablado

Wall tie: Ligadura de pared

ENGLISH-SPANISH GLOSSARY (Alphabetical Listing) *(cont.)*

Walls below grade:
Muros por debajo del
nivel de terreno

Want: Quiero

Warehouse:
Deposito, bodega

Warm water: Agua tibia

Warn: Advertir

Warning: Aviso

Warning signs:
Senales de peligro

Warning tape: Cinta
de cuidado

Warped: Torcido

Warrantee: Garantía

Wash: Lavar

Wash basin: Lavado

Washer: Rondana, lavadora

Washer and dryer:
Lavadora y secadora

Washroom: Baño

Watch: Tener cuidado

Water brush:
Brocha de agua

Water level: Nivel de agua

Water reducer:
Reductor de agua

Water table: Tabla del agua

Water truck:
Tanquero de agua

Waterline: Pipa de agua

Waterproofing: Impermeable

Wavy: Ondulado

Wax seal: Empaque de cera

We: Nosotros

Weak: Débil

Wear: Usar (ropa)

Weather: Tiempo
(atmosferico)

Wedge: Cuna, calzo

Wednesday: Miercoles

Weeds: Malas hierbas

Weld: Soldar

Welder: Soldador

Welder's vest: Chaleco
de soldadura

Welding apron: Delantal
de soldadura

Welding blanket:
Manta de soldadura

Welding gloves:
Guantes de soldadura

Welding lead: Cable
conductor de soldadura

Welding plate: Placa
de soldadura

Welding rod: Electrodo

Well (water): Aljibe,
pozo de agua

West: Oeste

Wet: Mojado

What: Que

Wheelbarrow: Carretilla, senda de concreto carretilla

When: Cuando

Where: Donde

Who: Quien

Why: Porque

Wide: Ancho

Width: Anchura

Winch: Torno

Window: Ventana

Window casing: Chambrana de ventana

Window frame: Marco de ventana

Windy: Ventoso

Wipe: Frotar ligeramente

Wire: Alambre

Wire brush: Cepillo de alambre

Wire connectors: Conectores de alambre, cable alambre conector

Wire feed welder: Soldadura de alimentación de alambre

Wire mesh: Tela metálica, malla de alambre

Wire reel: Carrete de alambre

With: Con

Withdrawal: Retirada

Withholding: Retener

Woman: Señorita, señora

Wood chisel: Escopio, formon, cincel

Wood flooring: Piso de Madera

Wood putty: Masilla para Madera

Wood shake (shingle): Teja de Madera, ripia

Wood shingle: Ripia

Wood stake: Estaca de Madera

Wood studs: Montantes, pies derechos de madera

Work (verb): Trabajo

Work light: Lampara de trabajo

Worker: Trabajador

Worm-drive saw: Sierra circular con tornillo sinfin

Wrench: Llave, llave inglea

Write (verb): Escribir

X-ray: Radiografía

Yards: Yardas

Yes: Si

You: Tu, usted

Your: Tus

Z clamp: Grapa en Z

Zero: Cero

NOTES – NOTAS

CHAPTER 21/CAPITULO 21
Español-Inglés

ESPAÑOL–INGLÉS GLOSARIO (LISTADO ALFABÉTICO)

911, Nueve uno uno:
911, Nine-one-one

A: To

A nivel: Flush

A tiempo: On time

Abarbetado: Span

Abertura: Opening

Abertura bosqueja:
Rough opening

Abertura de limpieza:
Cleanout (chimney)

Abra: Open

Abril: April

Abrir: Open (verb)

Acabado: Finished

Acabado ordinario:
Rough finished

Acabar: Finish

Acarrear: Haul

Acceso: Access

Accesorio: Fitting

Accidente: Accident

Aceite de formas: Form oil

Acera: Sidewalk, walkway

Acero: Steel

Acero de estructural:
Structural steel

Acero inoxidable:
Stainless steel

Acero para reforzardo:
Reinforcing steel

Acetileno: Acetylene

Acido: Acid

Acollador: Lanyard

Acoplamiento: Coupling

Acre: Acre

Acuerdo: Agreement

Acumulación: Buildup

Acunar: Shim

Adentro: Inside

Adhesiva: Glue, adhesive

Adios: Goodbye

Aditivos: Additives

Advertir: Warn

Afilador: Grinder

Afilador de mano:
Hand grinder

Afilar: Grind (metal)

Afuera: Outside

Agarre: Grip

**Agente de aeroclusión
de cemento:** Air-
entraining agent

Agente de vinculación: Bonding agent

Agosto: August

Agregados: Aggregates

Agregar: Add

Agregar agua: Add water

Agrietado: Cracked

Agua caliente: Hot water

Agua de desagüe: Runoff

Agua de exudación: Bleedwater

Agua fria: Cold water

Agua superficial: Surface water

Agua tibia: Warm water

Agujero: Hole

Agujero ciego: Knockout

Agusas negras: Sewage

Ahora: Now

Aire acondiciónado: Air conditioning

Aire libre: Open air

Aislamiento/la insulación: Insulation

Aislante plastico: Plastic insulator

Ajustar: Adjust

Alambre: Wire

Alambre de atadura: Tie wire

Alambre de la suspensión del techo: Ceiling suspension

Alambre de pollo: Chicken wire

Alambre testigo: Tracer

Alambres vivos: Live wires

Alarma: Alarm

Alarma de incendio: Fire alarm

Alarma de incendio manual: Manual pull station

Alarma de refuerzo: Backup alarm

Albañil: Mason

Albañileria, mamposteria: Masonry

Albardilla: Coping

Alcance: Scope

Alcantarilla: Culvert, sewer

Aldaba: Latch

Aleación para soldar: Brazing alloy

Alero: Eave

Alerta, vigilante: Alert

Alfombra: Carpet

Alfombreros: Carpet layers

Alicates de extensión: Channel-lock pliers

Alicates de presión, pinzas perras: Vise-grip pliers

Alicates, pinzas: Pliers

Alimentador: Feeder

Alinear: Align

Aljibe: Cistern

Aljibe, pozo de agua: Well (water)

Alla: There

Allanar: Smash

Almacenar: Store (verb)

Alquiler: Rental

Alquitran, brea, chapopote: Tar

Alternar: Alternate (verb)

Alto, Ruidoso: Loud

Altura: Height

Aluminio: Aluminum

Alzado: Façade

Alzar: Raise

Amarra, ligadura: Tie

Ambulancia: Ambulance

Amigo: Friend

Amoladora: Grinder

Amperias: Amperes

Ancho: Wide

Anchura: Width

Anciajes: Fasteners

Anclar: Anchor, affix

Andamiaje: Scaffolding

Andamio: Scaffold

Andamios: Scaffolding

Angulo de hierro: Angle iron

Angulo recto: Right angle

Anillo en D: D ring

Anivelado: Level

Anteojos: Goggles

Antes: Before

Antibiótico: Antibiotic

Anticorrosivo: Corrosion resistant

Antorcha: Cutting torch, torch

Anulación (remoción): Abatement

Apagador, interruptor de circuito: Circuit breaker

Apagar: Shut off, turn off

Aparejar: Rig

Apilar: Pile (verb), Stack (verb)

Apilar: Pound

Apisonar: Tamp

Aplicar: Apply

Apoyo: Bracing

Apoyo de pared: Wall brace

Aprendiz: Apprentice

Apretado: Tight

Apretar: Tighten

Apretar (una tuerca): Torque

Aprieta: Tight

Aprobado: Approved

Aprobar: Approve

Aquí: Here

Arandela: Grommet

Arbol: Tree

Arboles: Trees

Arbustos: Shrubs

Arcilla: Clay

Arco: Arc, bow

Area cargada: Loaded area

Area de cascos: Hard hat area

Area total: Gross area

Arena: Sand

Arena de silicona: Silica sand

Arena movediza: Quicksand

Arenisca: Sandstone

Arido: Dry

Armadura de cubierto: Truss

Armario de exhibición: Showcase

Armazon: Framework

Arnes de caida: Fall harness

Arnes de la caida: Fall harness

Arquitecto: Architect

Arranque de impacto: Impact wrench

Arregle: Fix (verb)

Arriostramiento: Bridging

Arriostramiento de vigueta: Joist bridging

Arroyo encintado: Curb and gutter

Artefactos eléctricos: Electrical fixture

Artefacto: Fixture

Asbesto: Asbestos

Aseguarse: Verify

Asegurado: Secured

Aseguranza: Insurance

Asegurar: Secure

Asentamiento: Slump

Asentamiento: Sprinkle

Aserrar: Saw (verb)

Asfalto: Asphalt

Aspero: Rough

Astilla: Sliver

Atadura: Ties

Atadura instantanea: Snap tie

Atar: Fasten, tie (verb)

Atarse: Tie off

Atarse seguramente: Tie off

Atorado: Jammed

Atornillar: Screw

Atrasado: Late

Atrio: Atrium

Autocierre: Self-closing

Auto-ignición: Self-ignition

Autoluminoso: Self-luminous

Aviso: Notice, warning

Aviso de proceder: Notice to proceed

Ayudante: Helper

Ayudar: Help

Azadón: Hoe

Azotado, jaharro: Plaster

Azotea: Roof

Azulejero: Tile setter

Azulejos de ceramicas: Ceramic tile

Bajante sanitaria: Soil stack

Bajar: Lower

Balance: Balance

Baldosas: Floor tiles

Banco: Bank

Banda de ligadura: Banding

Banda de ligadura: Bearing wall

Bandeja a tierra: Ground bar

Bandeja de carga: Hot bus bar

Bandeja de portacables: Cable tray

Bandeja neutral: Neutral bar

Bañera: Bathtub

Baño: Washroom

Banqueta de aserrado: Saw horse

Baranda, barra, carril: Railing

Barandal: Guard rail

Barandales: Handrails

Barra a tierra, vara a tierra: Ground rod

Barra de palanca: Pry bar

Barra de refuerzo, varilla: Rebar

Barra rellenadora: Filler rod

Barrer: Sweep

Barrera antihumo: Smoke barrier

Barreta: Bar

Barricada: Baricade

Base de pavimento: Road base

Basura: Garbage, trash

Basurero: Dump

Beneficiario: Payee

Bien: Good

Bisagras: Hinges

Blando, suave: Soft

Bloque: Block

Boca de riego: Hydrant

Bolsas: Bags

Bomba: Pump

Bomba de sumidero: Sump pump

ESPAÑOL–INGLÉS GLOSARIO (LISTADO ALFABÉTICO) *(cont.)*

Bombear: Pump (verb)

Bombero de concreto: Pump truck

Bono: Bond

Bono de ejecución: Performance bond

Borde agudo: Sharp edge

Borde de enlace: Bonding jumper

Borracho: Drunk

Botador de clavos: Nail set

Botas: Boots

Botas de goma: Rubber boots

Botas de goma: Screeds

Botas prectores de la punta del pie: Steel toe boots

Botella (oxígeno o acetileno): Bottle (oxygen, acetylene)

Botiquín: Medicine cabinet

Brazo: Bracket

Brecha: Gap

Brida: Stirrup

Brillar: Shine

Brilliante: Shiny

Broca, mecha: Drill bit

Brocha: Brush

Brocha de agua: Water brush

Brocha de pintura: Paint brush

Bronce: Brass

Buharda: Dormer

Buscador de montantes: Stud finder

Cabezal: Header

Cabio: Rafter

Cable a tierra: Ground wire

Cable ahogador: Choker

Cable conductor de soldadura: Welding lead

Cable conductor de tierra: Ground lead

Cable de alimentación: Feeder cable

Cable de anlace: Bonding conductor

Cable enterrado: Buried cable

Cable neutro: Neutral conductor

Cable principal: Main power cable

Cable principal neutro: Service entrance neutral

Cadena: Chain

Cadena de disposición: Positioning chain

Caerse: Fall

Caja de camión: Truck bed

Caja de conexiónes de empalme: Junction box

Caja de controles de regadera: Sprinkler control box

Caja de corte a ángulos: Mitre box

Caja de enchufe, Caja de tomacorriente: Outlet box

Caja de fusibles: Fuse box

Caja de trabajo: Job box

Cajon: Drawer

Cajones de aire comprimido: Caissons

Cal viva: Quicklime

Calafatear: Caulk

Calafeto: Caulking

Calcio: Calcium

Calcular: Calculate (verb)

Caldera: Boiler

Caldera: Kettle

Calefacción, ventilación y aire acondiciónado: HVAC

Calefacción: Heating

Calefactor, estufa: Heater

Caliber: Gauge (thickness)

Caliente: Hot

Caliza: Limestone

Calle: Street

Callejón: Alley

Calor: Heat

Calzada: Pedestrian walkway

Calzo: Wedge

Cambiar: Change (verb)

Cambiar de orden: Change order

Cambio: Change

Camilla: Batter boards

Caminar: Walk

Camioneta: Pick-up truck

Camino: Road

Caminos móviles: Moving walkway

Camión: Truck

Camión de descarga: Dump truck

Camión de tarima: Flat bed truck

Camisa, manga: Sleeve

Campana (cocina): Hood (kitchen)

Canal inclinado: Chute

Canaletas: Chase

Candente: Red hot

Cañeria, caño, tubo: Pipe, piping

Cantidad, suma: Amount

Capa: Ply

Capa de colada: Pour coat

Capa intermedia: Interlayment

Capataz: Foreman

ESPAÑOL–INGLÉS GLOSARIO (LISTADO ALFABÉTICO) (cont.)

Carga muerta: Dead load

Carga sísmica: Earthquake load

Cargador: Loader

Cargar (las pilas): Charge (batteries)

Cargas no balanceadas: Unbalanced loads

Cargas vivas: Live load

Carpas: Tents

Carpintero: Carpenter

Carreta: Cart

Carrete de alambre: Wire reel

Carrete entizado: Chalk box

Carrete retractable: Retractable reel

Carretilla: Wheelbarrow

Carril de montante: Stud track

Cartabon: Gusset

Casa: House

Cascara, cubierta: Shell

Casco: Hard hat

Catorce: Fourteen

Cáustico: Caustic

Cavar: Dig

Cavar un túnel: Tunnel

Celosia: Louver

Cemento: Cement

Cemento Portland: Portland cement

Cenizas volantes: Fly ash

Centro a centro: Center to center

Centro commercial: Mall

Cepillar: Brush

Cepillo: Plane

Cepillo automático: Jointer

Cepillo de alambre: Wire brush

Cepillo de mano: Hand brush

Cepillo de mano: Jointer plane

Cerca: Fence

Cerca, proximo: Near

Cero: Zero

Cerradura: Lock

Cerramiento: Enclosure

Cerrar: Lock (verb)

Cerrar (bajo de llave): Lock

Certificado de uso: Certificate of occupancy

Césped: Sod

Chaflán: Chamfer

Chaleco de soldadura: Welder's vest

Chalecos: Vests

Chambrana: Casing

ESPAÑOL–INGLÉS GLOSARIO (LISTADO ALFABÉTICO) *(cont.)*

Chambrana de ventana: Window casing

Chapeado: Plywood

Charco: Puddle

Cheque: Check

Chimenea: Chimney, fireplace

Chispas: Sparks

Cielo: Ceiling

Cien: One hundred

Cierre: Lockout/tagout

Cierre de tiro: Draft stop

Cincel: Chisel

Cinco: Five

Cincuenta: Fifty

Cinta de cuidado: Warning tape

Cinta de enmascarar: Masking tape

Cinta de medir: Tape measure

Cinta de precaución: Caution tape

Cinta pescadora: Fish tape

Circuito: Circuit

Ciudadanía: Citizenship

Ciudadano: Citizen

Clasificación: Rating

Clavado: Spiked

Clavador neumático: Nail gun

Clavar: Nail

Clave: Keystone

Clavo de cabeza perdida: Casing nail

Clavo especial para madera: Spike

Clavo oblicuo: Toe nail

Clavos: Nails

Clavos salientes: Protruding nails

Clavos sin cabeza: Finishing nails

Clínica: Clinic

Cloch: Clutch

Cobre: Copper

Cobre estirado en frio: Hard drawn copper

Cocina: Kitchen

Código: Code

Código de Incendios: Fire Code

Codo: Elbow

Colapso de agujero: Cave-in

Colgar: Hang

Colgar (puerta): Hang (door)

Collector de grasas: Grease trap

Colocar: Place

Colocar: Put

Columna: Column, queen post

Columna hidrante: Standpipe

Comba: Camber

Combado: Cambered

Combinación de cargas: Combination load

Combustible: Combustible

Como: How

Compactar: Compact

Competente: Competent

Complemento: Allowance

Comprar: Buy

Compressor de aire: Air compressor

Comprobar: Check (verb)

Compuerta: Hatch

Compuesto de curado: Curing compound

Comunicaciónes peligrosas: Hazardous communications

Comunicar: Communicate

Con: With

Con seguridad: Safely

Concreto: Concrete

Condominio residencial: Condominium

Conducir: Drive

Conducto: Duct

Conducto eléctrico: Raceway

Conducto portacables flexible: Flexible conduit

Conducto principal de gas: Gas main

Conducto, canal: Conduit

Conductor: Conductor

Conductos: Ducts

Conductos de humo: Flue

Conectar: Attach, connect

Conectar a tierra: Ground

Conector: Connector

Conector con tornillo: Screw connector

Conectores de alambre, cable alambre conector: Wire connectors

Conexión de tierra: Ground connection

Conexión, unión: Connection

Congelante: Freezing

Conocimiento de tráfico: Traffic awareness

Conciente: Conscious

Construcción apuntalada: Shored construction

Construcción no apuntalada: Unshored construction

Construir: Build

Contaminación: Contamination

Continuo: Continuous

Contraflujo: Backflow

Contrahuella: Riser (stair)

Contratar: Contract

Contratar, emplear: Hire, employ (verb)

Contratista: Contractor

Contrato: Contract

Convenir: Agree

Coordinar: Coordinate

Corcon: Bead

Cornisa: Cornice

Correcto: Correct, right

Correr (usar): Run (use)

Corriente CA: AC current

Corriente CC: DC current

Corrosivo: Corrosive

Cortado de césped, cortacésped: Lawnmower

Cortapernos: Boltcutters

Cortar: Cut (verb)

Corte: Cut

Corte limpio: Clean cut

Cortinas: Curtains

Cremaliera, tarima: Rack

Cremallera, barandal: Rail

Cresta: Ridge

Creyon de Madera: Keel

Cronograma de construcción: Construction schedule

Croquis: Lay out

Cristal roto: Broken glass

Cuadrado: Square

Cuadro de cortacircuito: Circuit breaker panel

Cuadro de cubierta de techo: Roofing square

Cuando: When

Cuarenta: Forty

Cuarto de almacenamiento: Storage room

Cuarto de baño: Bathroom

Cuarto de calderas: Boiler room

Cuarto de herramientas bajo llave: Tool lockup

Cuarto, sala: Room

Cuatro: Four

Cubeta: Bucket

Cubeta del trapeador: Mop bucket

Cubierta: Deck, decking

Cubierta de techo: Roof deck

Cubo: Bucket

Cubos de construcción: Block

ESPAÑOL–INGLÉS GLOSARIO (LISTADO ALFABÉTICO) *(cont.)*

Cubrejuntas, tapajuntas: Flashing

Cubrir: Coat, cover

Cuchara para cemento: Margin trowel

Cuchillo: Knife

Cuenta de banco: Bank account

Cuerda de extención: Extension cord

Cuerdas: Cords

Cuesta, Inclinación: Slope

Cumplir: Complete

Cuna: Wedge

Cuota: Quota

Curado con humedad: Moist curing

Curva: Curve

Dado fracciónario: Die/chaser nut

Dañado: Damaged

Dar: Give (verb)

De: From

De: Of

De altura: On grade

Debajo: Under

Deber: Owe

Débil: Weak

Debilitarse: Collapse (person)

Declive, inclinación: Incline

Descomponerse: Decompose

Defectuoso: Defective

Delantal de soldadura: Welding apron

Delgado: Thin

Diluyentes de pintura: Paint thinners

Demoler: Demolish

Departamento de construcción: Building department

Depósito: Deposit

Depósito de basurera: Land fill

Depósito, bodega: Warehouse

Derecho: Right, straight

Derecho de retención: Lien

Desaguar: Dewater

Desagüe: Drain, drainage

Desagüe de techo: Roof drain

Desarmador: Screw driver

Desarmador cruz: Philips screwdriver

Desarmar: Tear apart

Descanso de escaleras: Landing (stair)

Descarga, liberación: Release

Descargar: Dump, unload

Descartar: Throw away

Descolorido: Discolored

Desconectar: Disconnect switch

Descongelación: Thawing

Desechar: Scrap

Desempeño: Performance

Deshidratado: Dehydrated

Desigual: Uneven

Desinfectante: Disinfectant

Deslindar: Survey (verb)

Deslizar: Slip

Despacio: Slow

Despejar: Clear

Desperdicios: Rubbish

Desplazamiento, desvío: Offset

Despues: After

Destapacaños, sopapa: Plunger

Destino, tenencia: Occupancy

Destornillador, desarmador: Screwdriver

Destripador: Ripper

Destruir: Destroy

Detalles: Details

Detener: Hold

Determinar: Determine

Detras de: Behind

Día de paga: Payday

Día feriado: Holiday

Diámetro: Diameter

Dibujar: Draw

Dibujos: Drawings

Dibujos de taller: Shop drawings

Diciembre: December

Diecinueve: Nineteen

Dieciocho: Eighteen

Diecisiete: Seventeen

Dieciséis: Sixteen

Diesel: Diesel

Diez: Ten

Diferente: Different

Difícil: Difficult

Dimensiónado: Proportioned

Dimensiónes: Dimensions

Dinamitar: Dynamite

Dinero: Money

Dintel: Lintel

Dirección: Address

Direcciónes (Norte...): Directions (North...)

Disco: Circular saw blade

Disco de amoladora: Grinding wheel

Discrepar: Disagree

ESPAÑOL–INGLÉS GLOSARIO (LISTADO ALFABÉTICO) *(cont.)*

Dispensadores de papel: Paper dispensers

Disposición: Provision

Dispositivo de cierre automático: Automatic closing device

Después: Later

Distancia: Distance

Dividir: Divide

Doblador de varilla: Rebar bender

Dobladora portátil: Hickey bar

Doblar: Fold

Doblador de varilla: Bar bender

Doce: Twelve

Documentación: Documentation

Documentar: Document, record (verb)

Dolares: Dollars

Dolor: Pain

Domingo: Sunday

Donde: Where

Dos: Two

Dos por cuatro: Two by four

Dos por doce: Two by twelve

Draga: Dredge

Drenaje: Drainage

Drenaje para tormentas: Storm drain

Drenar: Drain (verb)

Ducha, regadera: Shower stall

Ductilidad: Ductility

Ducto: Chute

Ducto de basura: Rubbish cute

Dueño: Owner

Duro, firme: Hard

Ebanista: Cabinetmaker

Echar: Pour (verb)

Edificación insegura: Unsafe building

Edificio de departamentos: Apartment building

Edificio de gran altura: High-rise building

Efectivo: Cash

Eje: Hub

El: He

El amperaje: Amperage

Electricidad: Electricity

Electricista: Electrician

Electrodo: Welding rod

Elevación: Elevation

Elevador: Elevator

Elevar: Lift

Eliminación de polvo: Dust control

Ella: She

Ellos: Them

Embudo: Funnel

Emergencia: Emergency

Empanelado: Paneling

Empalmar: Splice (verb)

Empalme, translape, junta: Splice

Empaque de cera: Wax seal

Empezar: Start (verb)

Empujar: Push

En: In

En línea: On line

En punto (tiempo): Precisely (time)

En, sobre: On

Encarrilada: Track hoe

Encendedor: Lighter

Enchufar: Plug in

Enchufe, clavija: Receptacle, plug

Enchufe, tomacorriente: Electrical outlet

Encima: Over

Enclavamiento: Interlocking

Energía temporaria: Temporary power

Enero: January

Enfermo: Sick

Enfriar: Cool

Engrapadora: Stapler

Engrapadora automática: Stapling gun

Engrasar: Grease

Enjuagar: Rinse

Enjuagador de vidrio: Squeegee

Enlace, tirante: Link, linkage

Enlucido: Putty coat

Enparnar: Bolt (verb)

Enrasado: Furring, furred

Entablado: Wall sheathing

Entablado de techo: Roof sheathing

Entablonado: Planking

Entarimado de tejado: Roof sheathing

Entcontrar: Find (verb)

Enterrar: Bury (verb)

Entrada: Entry

Entrenamiento: Training

Entrenar: Train, teach

Entresuelo: Mezzanine

Erguir: Stand (verb)

Erigir: Erect

Erosión: Erosion

Escalera: Ladder

Escalera de extensión: Extension ladder

ESPAÑOL–INGLÉS GLOSARIO (LISTADO ALFABÉTICO) *(cont.)*

Escalera de tijera: Step ladder

Escaleras de caracol: Spiral stairs

Escalones: Stairs

Escape, extracción: Exhaust

Escoba: Broom

Escoba de concreto: Concrete broom

Escobón: Push broom

Escombro: Rubble

Escopio, formon, cincel: Wood chisel

Escoramiento: Shoring

Escoria: Slag

Escribir: Write (verb)

Escuadra: Carpenter's square, framing square

Escudo de cara: Face shield

Ese, esa: That

Eslinga: Sling

Espacio: Clearance

Espacio angosto: Crawl space

Espacio vacío: Void space

Espatula de masilla: Putty knife

Esperar: Wait

Esponja: Sponge

Esquilar: Clip

Esquinas: Corners

Estabilidad: Stability

Estabilizar: Stabilize

Estaca de bandera: Flag stake

Estaca de Madera: Wood stake

Estacas azules: Blue stakes

Estaciónimiento: Parking lot

Este: East

Este: This

Estera: Mat

Estéril: Sterile

Estimación: Estimate

Estimator: Estimator

Estribo: Pier

Estructura: Structure

Estuco: Stucco

Etiqueta: Tag

Etiquetar: Label

Evacuar: Evacuate

Examinar: Examine, inspect

Excavadora de arrastre: Dragline

Expansión: Swelling

Explosivo: Explosive

Exponer: Expose

Expuesto: Exposed

Extender: Extend

Exterior: Exterior

Extintor: Fire extinguisher

Extracción de los planos: Take off

Exudarse (agua): Bleed (water)

Fábrica: Factory

Fabricado: Fabricated

Fabricante: Manufacturer

Fabricar: Fabricate, manufacture

Fácil: Easy

Falso plafon, cieloraso suspendido: Suspended ceiling

Familia: Family

Fangoso (nieve): Slushy

Fangoso (tierra): Muddy

Faso: Phase

Fase 1, fase 2, fase 3: A phase, B phase, C phase

Febrero: February

Fecha: Date

Fertilizante, abono: Fertilizer

Fibra transversal: Cross grain

Fieltro: Felt

Filoso: Sharp

Finos de agregado: Aggregate, fines

Firma: Signature

Fleje: Strap

Flejes: Strapping

Flojo: Loose

Florescente: Fluorescent

Flota de esponga: Sponge float

Flota de neoprene: Neoprene float

Foco: Light bulb

Fontanero: Plumber

Forma: Form

Forma de zapata de cimentación: Footing form

Formar una bloqueda: Block out

Formas de borde: Edge forms

Formas de escalera: Stair forms

Formas de estribo y columna: Pier and column forms

Formas de losa: Slab forms

Formas de viga: Beam forms

Formas para pieza plana: Screed rod

Forrado, revestimiento: Siding

Fosa septica: Septic tank

Fosforescente: Red hot

Fosforos: Matches

Frágil: Fragile

Fraguarse: Set up

Fregadero: Sink

Fregar: Scrub

Freno de puerta: Door closer

Frente, delante: Front

Frecuencia: Frequency

Fresadora, contorneador: Router

Fresco/verde: Fresh/green

Frigorífico: Walk-in cooler

Frío: Cold

Friso, alfarje: Wainscoting

Frotar: Rub

Frotar ligeramente: Wipe

Fuego: Fire

Fuente de alimentación: Power supply

Fuera de: Off

Fuerte: Strong

Fugs, gotera: Leak

Fundación: Foundation

Fundente: Flux

Fundente para soldar: Brazing flux

Fusible: Fuse

Fusible de cartucho: Cartridge fuse

Fusible de rosca: Plug fuse

Gabinetes: Cabinets

Galvanizado: Galvanized

Gancho: Hook

Ganchos: Hangers

Garantía: Warrantee

Gas natural: Natural gas

Gases: Fumes

Gasolina: Gasoline

Gato: Jack

Gavetas: Lockers

Generador: Generator

Gerente: Manager

Gimnasio: Gymnasium

Golpear (una estaca): Drive (a stake)

Gotera: Gutter

Gradas: Steps

Grado: Grade

Grados (ángulos): Degrees (angle)

Grande: Large

Grapa en C: C clamp

Grapa en Z: Z clamp

Grapa, abrazadera: Clamp

Grasiento: Greasy

Grava: Gravel

Gravita: Pea gravel

Grifo de manguera: Hose bibb, sill cock

Grillete: Shackle

Grua: Crane

Gualdrapear: Rack

Guantes: Gloves

ESPAÑOL–INGLÉS GLOSARIO (LISTADO ALFABÉTICO) *(cont.)*

Guantes de cuero: Leather gloves

Guantes de goma: Rubber gloves

Guantes de soldadura: Welding gloves

Guardado de proximidad: Perimeter guard

Guardar: Keep

Guarnición: Curb

Habitación, dormitorio: Bedroom

Hablar: Speak (verb)

Hacer cumplir: Enforce

Hacer disposiciónes: Lay out

Hacha: Axe

Halar: Drag

Hastial: Gable

Hecho: Done

Helada: Frost

Herido: Hurt

Herramientas: Tools

Herramientas de mano: Hand tools

Hidreción: Hydration

Hielo: Ice

Hilo: Thread

Hoja: Blade

Hola: Hello

Hombre: Man

Hora: Hour

Horario: Schedule

Horarios: Schedules

Hormigón preesforzado, hormigonprecargado: Prestressed concrete

Horno: Furnace

Hospital: Hospital

Hoy: Today

Hoyo, pozo: Pit

Hulla: Coal

Humedad: Moisture

Humedo: Damp

Humo: Smoke

Hundirse: Sink

Hígiene: Sanitation

Ignifugación: Fire proofing

Ilegal: Illegal

Ileso, seguro: Safe

Iluminación: Floodlight

Iluvioso: Rainy

Impedencia: Impedance

Impelar: Ram

Impermeable: Waterproofing

Impermeables al humo: Smoke-tight

Imprimido: Primed

Imprimidor: Primer

Impuestos: Taxes

Impuestos estatales: State taxes

Impuestos federales: Federal taxes

Impuestos locales: Local taxes

Incompetente: Incompetent

Incorrecto: Incorrect, wrong

Indicadores: Gauges

Inflamable: Flamable

Ingeniero: Engineer

Ingeniero geotecnico: Geotechnical engineer

Ingeniero de tierras: Soil engineer

Inodoro, sanitario: Toilet

Inquilino: Tenant

Insertar: Insert

Insolación: Heat stroke

Inspección: Inspection

Inspector: Inspector

Inspector de obras: Building inspector

Instalación de hidráulicas y sanitarias, plomeria: Plumbing

Instalación en obra negra, instalación de obra gruesa: Rough-in

Instalaciónes esenciales: Essential facilities

Instalar: Install

Interceptor de grasas: Grease interceptor

Interior: Interior

Interprete: Interpreter

Interruptor automático bipolar: Double pole breaker

Interruptor automático principal: Main breaker

Interruptor automático unipolar: Single pole breaker

Interruptor de vacío: Vacuum breaker

Interruptor, apagador: Switch

Interuptor fusible de seguridad a tierra: Ground fault circuit Interruptor (GFCI)

Intramado: Framing

Inyección contra el tétanos: Tetanus shot

Ir: Go (verb)

Jabalcon: Brace

Jabón: Soap

Jaboncillo: Soap stone

Jacena exterior, timpano: Spandrel

Jalar: Pull

Jamba: Jamb

Jamba de puerta: Door jamb

Jefe: Employer, boss

Jueves: Thursday

Julio: July

Junce: Junk

Junio: June

Junta: Joint

Junta de cemento: Spackle

Junta de control: Control joint

Junta de expansión: Expansion joint

Junta de tope: Butt joint

Junta en V: V joint

La falta: Missing

La mañana: Morning

Lado: Mud

Ladrillo: Brick

Ladrillo de cara: Face brick

Ladrillo de fuego: Firebrick

Ladrillos para frentes: Facing brick

Lámina de cobre: Sheet copper

Lámina metálica, chapa metálica: Sheet metal

Lámpara: Lamp

Lámpara de trabajo: Work light

Lápiz: Pencil

Larguero: Waler

Larguero central: Mullion (door)

Largueros: Runners

Laser: Laser

Lastre: Ballast

Lata, chapa, estano: Tin

Lavado: Bathroom sink, wash basin

Lavadora y secadora: Washer and dryer

Lavar: Wash

Lazadas: Loop

Lechada: Grout

Leer: read (verb)

Legal: Legal

Lejos: Far

Lentamente: Slowly

Lento: Slowly

Lesión: Injury

Levantar: Lift (verb), raise

Ligadura de pared: Wall tie

Ligadura plana: Flat ties

Ligaduras de larguero: Waler loops

Ligaduras cono de pared: Cone ties

Lija: Sand paper

Lijadora: Sander

Lijar: Sand (verb)

Lima: File

Lima: Hip

Limite: Boundary

ESPAÑOL–INGLÉS GLOSARIO (LISTADO ALFABÉTICO) *(cont.)*

Limpiar: Clean (verb)

Limpio: Clean

Líneas de energía: Power lines

Líneas de rejilla: Grid lines

Líneas de servicios publicos: Utility lines

Líneas subterráneas: Underground lines

Línea de guía: Tag line

Línea de hilo: String line

Línea de la pared: Wall line

Línea de limite de la propiedad: Property boundary line

Línea de propiedad: Property line

Linterna: Flashlight

Líquido inflammable: Flammable liquid

Líquido refrigerante: Coolant

Liso: Slick, smooth

Listo: Ready

Liston: Lath

Liviano: Light

Llamar: Call

Llana: Square trowel

Llana de emplastar: Plastering trowel

Llana mecánica: Power trowel

Llave: Key, wrench

Llave de agua: Faucet

Llave de asiento de valvula: Valve-seat wrench

Llave de cadena: Chain wrench

Llave de cimentación: Keyway

Llave de cola: Spud wrench

Llave de correa, llave de cincho: Strap wrench

Llave de flujo: Ball valve

Llave de tubo: Pipe wrench

Llave de tuercas, llave adjustable: Crescent wrench

Llave dinamometrica: Torque wrench

Llave francesa: Adjustable wrench

Llave inglesa: Wrench

Llave pico de gansa: Basin wrench

Llave, grifo, canilla: Spigot

Llenar: Fill

Lo siento: I'm sorry

Local, sitio: Premises

Localizado: Located

Localizar: Locate

Lona: Tarp

Longitud: Length

Losa: Slab

Lubricar: Lubricate

Lucernario: Skylight

Luces: Lights

Lugar de la obra: Job site

Lunes: Monday

Machihembrado: Tongue and groove

Madera de construcción: Lumber

Madera de construcción tratada: Treated lumber

Madera de estaciónal: Seasoned wood

Madera de secoya: Redwood

Madera elaborada: Graded lumber

Madera laminada para formar: Form ply

Magnesio: Magnesium

Malas hierbas: Weeds

Malla de alambre: Wire mesh

Malo: Bad

Mampostería reforzada: Reinforced masonry

Mañana: Tomorrow

Mandil, delantal: Carpenter's apron

Mangos de herramientas (tubos): Tool handles (poles)

Manguera: Hose

Manguera de jardín: Garden hose

Manija: Handle

Manipular: Handle (verb)

Mansardara: Mansard roof

Manta de soldadura: Welding blanket

Mantas: Blankets

Mantenimiento: Maintenance

Mapa topográfico: Topographic map

Máquina: Machine

Máquina niveladora: Grader

Maquinista: Machinist

Marcador de juntas de mano: Hand jointer

Marcar: Mark

Marcar (teléfono): Dial (phone)

Marco de puerta: Door frame

Marco de ventana: Window frame

Marco, estructura: Frame

Marro, mazo: Sledgehammer

Martes: Tuesday

Martillito: Beater

Martillo: Hammer

Martillo chivo: Claw hammer

Martillo de clavos: Nail hammer

Martillo de escoria: Slag hammer

Martillo neumático: Jackhammer

Marzo: March

Mas: More

Mascara, careta: Mask

Mascarilla para polvo: Dust mask

Masilla para Madera: Wood putty

Mastique: Mastic

Mastique de cal: Lime putty

Mayo: May

Mazo: Mallet

Mecánico: Mechanic

Medicación: Medicine

Medicare: Medicare

Medidor: Meter

Medidores, Metros: Gauges

Medio: Middle

Medios de salida: Means of egress

Medir: Measure

Mejor: Better

Menos: Less

Metal ornamental: Ornamental metal

Metro pernos: Bolt

Metro, medidor: Meter (measuring device)

Mezcladora de concreto sobre camión: Concrete mixing truck

Mezclar: Mix

Mi: My

Miembros: Limbs

Miercoles: Wednesday

Mil: One thousand

Minuto: Minute

Mirar: Look (verb)

Mismo, igual: Same

Modificación: Modification

Modificar: Modify

Moho: Rust

Mojado: Wet

Molcajete, mortero: Mortar

Moldura: Molding

Molduras: Trim

Moler: Grind

Monometro: Gauge (instrument)

Montengase fuera: Keep out

Montantes cojos: Cripple stud

Montantes, pies derechos de acera: Steel studs

Montantes, pies derechos de madera: Wood studs

Mordaza tiradora de alambre: Come-along

Morsa: Vise

Mortal: Lethal

Motivado: Motivated

Motor: Motor

Mover: Move

Mucho: Much

Muchos: Many

Mueble sanitario: Plumbing appliance

Muesca, ranura: Rabbet

Muescado: Nicked

Muestra de pernos: Bolt patterns

Mueva: Move (verb), mover

Muro con montantes: Stud wall

Muro cortante: Sheer wall

Muro de fundación: Foundation wall

Muro exterior: Exterior wall

Muro hueco: Cavity wall

Muros por debajo del nivel de terreno: Walls below grade

Navaja de utilidad: Utility knife

Necesitar: Need (verb)

Neutro: Neutral

Nieve: Snow

Niños: Children

Nivel: Level

Nivel de agua: Water level

Nivelar: Grade

No: No

Noche: Night, evening

Nombre: Name

Nombre de banco: Bank name

Norte: North

Nosotros: Us, we

Notas: Notes

Noventa: Ninety

Noviembre: November

Nudo de Madera: Knot

Nueve: Nine

Número de cheque: Check number

Número de cuenta: Account number

Número de ocupantes: Occupant load

Número de seguro social: Social security number

Número de télefono: Phone number

ESPAÑOL–INGLÉS GLOSARIO (LISTADO ALFABÉTICO) (cont.)

Nunca: Never

O: Or

Ochenta: Eighty

Ocho: Eight

Octubre: October

Oeste: West

Oferta: Bid

Oficial: Journeyman

Oficina: Office

OK: OK

Omposta: Fascia

Once: Eleven

Ondulado: Wavy

Operador de grua:
Crane operator

Ordenar: Order

Orillero: Edger

Oscilación: Swing

Oscilar: Swing (verb)

Oscuro: Dark

Otra vez: Again

Oxígeno: Oxygen

Pago: Payment

Paisajista: Landscaper

Pala: Shovel

Palanca: Level

Palanca de cola: Sleever bar

Paleta de albañil:
Mason's trowel

Paleta de relleno:
Joint-filler trowel

Palo de grado: Grade stick

Panaleado: Honeycombed

Panel: Board, panel

Panel de yeso:
Gypsum board

Paneles acusticos:
Acoustical panels

Papel de brea: Tar paper

Papel Kraft: Kraft paper

Paralelo: Parallel

Parar: Stand (verb), stop

Parchar: Patch

Pared: Wall

Pared de parapeto:
Parapet wall

Partición: Partition

Pasador de varilla: Dowel

Pasamanos ornamental:
Ornamental railing

Pasaporte: Passport

Pasar la aspiradora:
Vacuum (verb)

Pasillo: Corridor, aisle

Pasta de muro:
Joint compound

Pasta de muro:
Taping compound

Pata de chiva: Cat's paw

Pedrón: Boulder

Pegado: Stuck

Pegamento: Adhesive, glue

Pegamento de construcción: Construction adhesive

Pegar: Hit

Pegar: Stick

Peligro: Danger, hazard

Peligroso: Dangerous, hazardous

Pendiente del techo: Pitch

Pendiente, talud: Slope

Penetración: Penetration

Pensión: Pension

Pequeño: Small

Percutor: Striker

Perfil a sombrero: Hat channel

Perfil en U: U channel

Perforación: Drilling

Permiso: Permit

Perno: Bolt

Perno de aliniación: Bull pin

Perno de anclaje: Anchor bolt

Perno de expansión: Expansion bolt

Pernos: Pins

Pero: But

Perpendicular: Perpendicular

Persona que da señales: Flagger

Personal autorizado solamente: Authorized personnel only

Pesado: Heavy

Picar: Chip

Piedra: Stone

Piedra de mano: Hand stone

Piedra moldeada: Cast stone

Pies cúbicos: Cubic feet

Pieza plana: Flat work

Pila: Battery

Pila: Pile

Pila de metal: Metal pile

Pilego, chapa, plancha: Sheet

Pilistra: Pilaster

Pino: Pine

Pintar: Paint

Pintor: Painter

Pintura de marcar: Paint mark

Pintura rociada: Spray paint

Pinzas pequeñas: Tweezers

Pipa: Pipe

Pipa de agua: Waterline

Pipas: Pipes

Pirca: Drywall

Pirca: Interior finish

Piscina (de natación), alberca: Swimming pool

Piso: Floor

Piso de Madera: Wood flooring

Pisón: Tamper

Pistola de calafeto: Caulking gun

Pistola de cartuchos para fijación: Powder nailer

Pistola descarmadora: Screw gun

Placa: Plate, top plate

Placa de anclaje: Embed plate

Placa de base: Base plate, sole plate

Placa de interruptor: Switch plate

Placa de soldadura: Welding plate

Placa embutida: Embed plate

Placa superior: Top plate

Plana: Float

Plana grande: Bullfloat

Plana, flatacho: Darby

Planear: Plan

Plano: Flat

Plano de plomería: Plumbing plan

Plano eléctrico: Electrical plan

Plano mecánico: Mechanical plan

Planos arquitectonicos: Architectural plans

Planos estructurales: Structural plans

Planos de sitio: Site plans

Plantador: Planter

Plantar: Plant (verb)

Plantas: Plants

Plantillas: Templates

Plasticizador: Plasticizer

Plastico: Plastic

Plataforma: Floor deck

Plataforma hidráulica: Scissor lift

Plataforma metálica: Metal deck

Pleno, camara de distribución de aire: Plenum

Plomada: Plumb

Plomear: Plumb (verb)

Plomo: Plumb bob

Pluma: Pen

Pocos: A few

Póliza de seguridad: Safety policy

Polvo: Dust

Polvoriento: Dusty

Poner: Put on

Poner (alfombra):
Lay (carpet)

Poner, colocar: Put (verb)

Poquito: Little

Por: For

Por que: Because

Porque: Why

Portal: Doorway

Portátil: Portable

Posiciónar: Position

Poste: Pole, post

Poste principal: King post

Potencia: Power

**Pozo de confluencia,
boca de inspección,
boca de acceso, pozo
de entrada:** Manhole

Precalentar: Preheat

Precaución: Caution

Precio: Price

Precio de modulo: Unit price

Prender: Turn on

Prensar: Press

Preparación: Preparation

Presión: Pressure

Prestamo: Loan

Primera: First

Principal, matriz: Main

Privado: Private

Probar: Test

Problema: Problem

Producto químico: Chemical

Profundidad: Depth

Promesa: Promise

Pronto: Quickly

Propano: Propane

Propiedad: Property

Proporción, dimensiónar:
Proportion

Protección: Protection

Protección de cabeza:
Head protection

Protección de caida:
Fall protection

Protección de ojos:
Eye protection

Protección contra termitas:
Termite protection

**Protector de serrucho
eléctrico:** Saw guard

Provisiónal: Temporary

Puerta: Door

Puerta de cerco: Gate

Puerta de salida: Exit door

Puerta giratoria:
Revolving door

Puerta pivotante:
Swinging door

Puerta trasera de camión: Truck tailgate

Puertas mecanicas: Power doors

Puesto: Set

Pulir: Burn in, polish

Punta: Point (shape)

Puntal: Kicker, strut

Puntales: Shoring

Punto alto: High spot

Punto bajo: Low spot

Punto de anclaje: Anchorage point

Punto de pellizco: Pinch point

Punto de referencia: Bench mark

Punzones: Punches

Que: What

Quebrar: Break

Quedarse fuera: Stay off

Quemadura de relampago: Flash burns

Quemaduras: Burns

Quemaduras de cemento: Concrete burns

Quemar: Burn

Quien: Who

Quiero: Want

Quince: Fifteen

Quiosco: Kiosk

Quitar: Remove

Quitar el polvo: Dust

Radio: Radius

Radiografía: X-ray

Rajadas: Cracks

Ramal: Branch

Ramal lateral: Lateral (pipe)

Rampa: Ramp

Ranura: Mortise

Ranuradora: Groover

Rapidamente: Quickly

Rapido: Fast, quickly

Rasante: Ground elevation

Raspar: Scrape

Rastillar: Rake

Rastrillo: Rake

Rayero de tiza: Chalk line

Rebaba: Burr

Rechazado: Rejected

Recibo del concreto: Vibrator

Reciclar: Recycle

Recinto: Shaft

Recinto de escaleras: Stairwell

Recinto de ventilación: Vent shaft

Recogedor de polvo: Dust pan

Recoger: Collect

ESPAÑOL–INGLÉS GLOSARIO (LISTADO ALFABÉTICO) *(cont.)*

Recoger: Pick up, scoop

Recubrimiento: Lining

Recumbrimiento: Covering

Recuperar: Recover

Reductor de agua: Water reducer

Reemplazar: Replace

Reflectores: Reflectors

Reformatario: Reformatory

Reforza: Reinforce

Reforzar: Reinforce

Refuerzo: Reinforcement

Regadera: Showerhead

Regaderas: Sprinklers

Regaderas de fuego: Fire sprinklers

Región, tramo: Region

Registrar: Record

Registro: Cleanout

Reglas: Rules

Regleta: Reglet

Regulador: Damper, regulator

Rejilla: Grille

Relación, cociente: Ratio

Relación, razón: Rate

Relleno: Fill, backfill

Relleno sin consolidar: Unbalanced fill

Remanche: Rivet

Remonte: Uplift (wind)

Rentar: rent (verb)

Reparación: Repair

Reparar: Repair (verb)

Repisa: Shelf

Repite: Repeat (verb)

Reponer: Replace

Reporte: Report

Reporte del día: Daily report

Reportaje de trabajo: Job report

Requerir: Require

Resbaladizo: Slippery

Residencias para estudiantes: Dormitory

Resistencia: Resistance

Resorte: Spring

Respiradero de bajante: Stack vent

Respiradero de sumidero: Sump vent

Respiradero matriz: Main vent

Respiradero vertical: Vent stack

Respirador: Respirator

Responsabilidad: Responsibility

Responsable: Responsible

Resquicio: Undercut

Retardador: Retardant

Retener: Withholding

Retirada: Withdrawal

Retiro: Setback

Retrasar: Delay

Retroajuste: Retrofit

Retroexcavadora: Backhoe

Reverso roscado: Reverse threaded

Revestimiento: Veneer

Revestimiento de aluminio: Aluminum siding

Revestimiento de chimenea: Chimney liner

Revestimiento de tablas con traslape: Lap siding

Revestimiento de techo, cubierta de techo: Roof covering

Revestimiento de vinilo: Vinyl siding

Revestimiento, laminado: Sheeting

Revestimientos para pisos: Flooring

Revoque, enclucido: Plastering

Riostras: Bracing

Ripia: Wood shingle

Roble: Oak

Roca de fondo: Bedrock

Roca, piedra: Rock

Rociador: Spray bottle, sprayer, sprinkler, sprinkler head

Rociar: Spray

Rodapie: Baseboard

Rodar: Roll

Rodilleras: Knee pads

Rodillo: Roller

Rodillo de pintura: Paint roller

Romper: Break

Romperse: Collapse (structure)

Rondana: Washer

Rotamartillo: Hammer drill

Roto, averiado: Broken

Ruina: Debris

Sabado: Saturday

Saber: Know (verb)

Sacudida de eléctrica: Shock

Sala de clase: Classroom

Salario: Salary

Salida: Egress, exit

Saliente: Protruding

Salvar: Rescue, salvage

Sanidad: Sanitation

Sangrar: Bleed

Sanitario: Sanitary

Sano: Healthy

Sección: Section, slump

Seco: Dry

Secretario, secretaria: Secretary

Seguramente: Securely

Seguridad pública, protección al público: Public safety

Seguro: Insurance

Seguro de salud: Health insurance

Seis: Six

Sellador: Sealant

Sellar: Seal

Sello de trampa hidráulica: Trap seal

Sembrar: Seed

Señal de mano: Hand signal

Señalar: Point (verb)

Señalar con la mano: Signal

Señales de mano de grua: Crane hand signals

Señales de peligro: Warning signs

Senda de concreto carretilla: Wheelbarrow

Señorita, señora: Woman

Sensor de humo: Smoke detector

Seperador: Spreader

Septiembre: September

Serie: Series

Serpiente: Snake

Serrucho: Saw

Serrucho circular: Circular saw

Serrucho de mesa: Table saw

Serrucho de mano: Hand saw

Serrucho guillotina: Radial arm saw

Serrucho tajadero: Chop saw

Servicios publicos: Public utilities

Sesenta: Sixty

Setenta: Seventy

Si: Yes

Siempre: Always

Sierra alternativa: Reciprocating saw

Sierra circular con tornillo sinfin: Worm-drive saw

Sierra circular de mano: Circular saw

Sierra de arco para metal: Hack saw

Sierra de cadena: Chain saw

Sierra de mano: Hand saw

Sierra de retroceso para ingletes: Mitre saw

Sierra de vaiven: Jigsaw

Sierra para cortar: Cut-off saw

Siete: Seven

Sifón: Trap

Simbolos: Symbols

Sistema de alcantarilla sanitario: Sanitary sewer

Sistema de columna hidrante: Standpipe system

Sistema de comunicación: Communication system

Sistema de evacuación: Venting system

Sistema de rociadores: Sprinkler system

Sistema de seguridad: Security system

Sistema de televisión: Television system

Sistema de ventilación: Vent system

Sitio de la caldera: Boiler room

Sobra: Extra

Sobradillo: Penthouse

Sofito: Soffit

Solamente: Only

Soldador: Welder

Soldadura a charco: Puddle weld

Soldadura a pano: Flush weld

Soldadura a solape: Lap weld

Soldadura a tope: Butt weld

Soldadura de arco: Arc welding

Soldadura de alimentación de alambre: Wire feed welder

Soldadura de encima: Overhead weld

Soldadura de plano: Flat weld

Soldadura ortogonal: Fillet weld

Soldadura vertical: Vertical weld

Soldar: Weld

Soldar en fuerte: Braze

Solera: Soleplate

Solera inferior: Sill plate

Solido: Solid

Soltar: Release (verb)

Someter: Submit

Sopladora: Blower

Soplete: Soldering torch

Soporte: Backing

Soporte de vigueta: Joist hanger

Sótano: Basement

Subcontratista: Subcontractor

Subsuelo: Subfloor, subgrade

Subterráneo: Underground

Sucio: Dirty

Suelo: Soil

Suelo subyacente: Underlayment

Suficiente: Enough, sufficient

Sujetador: Restraint

Sujetar: Fasten

Sujetar, formar: Frame

Sumidero: Sump

Superintendente: Superintendent

Supervisor, supervisora: Supervisor

Suplente: Alternate, alternative

Sur: South

T, injerto: Tee

Tabilla: Splint

Tabique movible: Moving partition

Tabique plegable: Folding partition

Tabique portátil: Portable partition

Tabla de cumbrera: Ridge board

Tabla de pie: Toe board

Tabla del agua: Water table

Tablero de lámina: Decking

Tablero prensado: Particle board

Tablones: Planks

Taladrar, agujererar: Drill (verb)

Taladro: Drill

Taladro de angulo recto: Right angle drill

Taladro de mano: Bit and brace

Taladro eléctrico: Electric drill

Taller: Shop

Tambo de basura: Dumpster

Tandem: Tandem

Tanquero de agua: Water truck

Tanques: Tanks

Tapa de acceso: Access cover

Tapadera de barilla: Rebar caps

Tapanco: Attic

Tapas: Caps

Tapas de botella: Bottle caps

ESPAÑOL–INGLÉS GLOSARIO (LISTADO ALFABÉTICO) *(cont.)*

Tapas de varilla: Rebar caps

Tapones del oído: Ear plugs

Tarde: Afternoon

Tarde: Late

Tarjeta de residencia: Green card

Techero: Roofer

Técnico: Technician

Techo: Roof

Techo a dos agues: Gable roof

Techo de cuatro agues: Hip roof

Techo plano: Flat roof

Teja: Roof tile, shingle

Teja de asfalto: Asphalt shingle

Teja de Madera, ripia: Wood shake (shingle)

Teja de pizarro: Slate shingle

Tejas de cielo: Ceiling tiles

Tejas para cumbrera: Ridge tile

Tela: Cloth

Tela asfaltica, felpa: Roofing felt

Tela metálica: Wire mesh

Teléfono: Telephone

Temprano: Early

Tener: Have (verb)

Tener cuidado: Watch

Tensión: Tension

Tensor: Turn buckle

Teodolito: Theodolite

Terminal, sin salida: Dead end

Termostato: Thermostat

Terrapien: Earth work, embankment

Terreno inestable: Unstable ground

Terrón, gelba: Clod

Textura: Texture

Tiempo (atmosferico): Weather

Tiempo fraguado: Setting time

Tienda: Store

Tierra: Dirt, earth

Tierra vegetal: Topsoil

Tijeras de varilla: Rebar shears

Tijeras para metal: Sheet metal shears

Timbre: Door bell

Tipo de suelo: Soil type

Tirante de formado: Cleat

Tirar (una línea): Shoot (a line)

Toldo: Canopy

Toldos: Awnings

Tomacorriente para estufa, enchufe para estufa: Range outlet

Tomar: Take (verb)

Tope: Door stop

Tope antifuego: Fire stop

Tope de forma: Bulkhead

Topografar: Survey

Topografo: Surveyor

Torcido: Distorted, warped

Tornillo: Screw

Tornillo de banco: Vise

Tornillos: Screws

Tornillos autoperforantes: Self-drilling screws

Tornillos autorroscantes: Self-tapping screws

Torno: Winch

Torno de banco: Vice bench

Torre de refrigerado: Cooling tower

Torreno, lote: Lot

Trabajador: Worker

Trabajador del conducto: Duct worker

Trabajo: Job, work (verb)

Trabajo de arriba: Overhead work

Trabajo de cooperación: Teamwork

Tracción: Traction

Traer: Bring

Tragaluz, claraboya: Skylight

Tranquilo, silencio: Quiet

Transceptor portátil: Walkie talkie

Transformador: Transformer

Transito: Transit

Transito de computadora: Total station

Transplantar: Transplant

Transportador mecánico: Conveyor

Transportar: Haul

Trapeador: Mop

Trapeador de polvo: Dust mop

Trapear: Mop (verb)

Trapos: Rags

Traslapar: Overlap (verb)

Traslapo: Overlap

Traspalar: Shovel (verb)

Travado: Stuck

Travesano: Blocking

Trazador de metal: Metal scribe

Trazar: Plot

Trece: Thirteen

Treinta: Thirty

Tres: Three

ESPAÑOL–INGLÉS GLOSARIO (LISTADO ALFABÉTICO) *(cont.)*

Trinquete a polea: Come-along

Tripode: Tripod

Tu, usted: You

Tubo vertical: Vertical pipe

Tuberia: Pipeline

Tuberia vertical bajante: Stack

Tubo bajante de aguas negras: Soil pipe

Tubo de bajada: Leader (pipe)

Tubo vertical: Riser (pipe)

Tuerca: Nut

Tuercas: Nuts

Tumbar: Knock over

Túnel: Tunnel

Tus: Your

Ultima: Last

Umbral: Door sill, threshold

Umbral bosquejo: Rough sill

Un cuarto: One quarter

Un millón: One million

Una vez: Once

Unidad de vivienda: Dwelling unit

Uniforme: Even, same

Unión, sindicato: Union

Uno: One

Urinario: Urinal

Usar (ropa): Wear

Vaciar: Empty

Vacío, aspiradora: Vacuum

Valor: Value

Valuación: Valuation

Valvula: Valve

Valvula de alivio, llave de alivio: Relief valve

Valvula de cierre: Cut-off valve

Valvula de contraflujo: Check valve

Valvula de cubo: Hub valve

Valvula de flotador: Ball cock

Valvula de llave: Key valve

Valvula de purga: Bleeder valve

Valvulas de cierre: Shutoff valve

Valvulas de regulación: Regulator valves

Vapor: Steam

Varilla de esquina: Corner bar

Varillas adiciónales: Backup bar

Varillas longitudinales y transversales: Longitudinal and transverse bars

Ve: Go (you)

Veinte: Twenty

Venas: Veins

Vendaje: Bandage

Vender: Sell

Veneno: Poison

Venir: Come (verb)

Ventana: Window

Ventilador: Fan

Ventilador de extacción: Exhaust fan

Ventilar: Ventilate

Ventilar, evacuar: vent (verb)

Ventoso: Windy

Verificar: Verify

Vertice: Crown

Vesiculación: Blistering

Vestíbulo: Hallway, vestibule

Veta superficial: Face grain

Via de acceso: Driveway

Vibrar: Vibrate

Vidriado: Glazing

Vidriado de seguridad: Safety glazing

Vidrio: Glass

Viejo: Old

Viernes: Friday

Viga: Beam

Viga de alma llena: Plate girder

Viga de carga: Load-bearing joist

Viga de fundación: Grade beam

Viga del borde: Rim beam

Viga doble T: I beam

Viga principal: Floor girder, girder

Vigueta: Floor joist, joist

Vigueta de cielo: Ceiling joist

Vigueta doble: Double joist

Visa: Visa

Vista de arriba: Overhead view

Vista de costado: End view

Vista de frente: Front view

Vitrina: Show window

Vivienda: Dwelling

Voladizo: Cantilever

Voladizo, vuelo, alero: Overhang

Volar: Blast

Volquete: Dump truck

Voltaje: Voltage

Voltios: Volts

Y: And

Yardas: Yards

Yeseros: Sheetrockers

Yeso: Gypsum

Yo: I

ESPAÑOL–INGLÉS GLOSARIO (LISTADO ALFABÉTICO) *(cont.)*

Zanca de escalera:
Stair stringer

Zanja: Trench/ditch

Zapapico: Pick axe

Zapata: Footing

Zapata de cimentación:
Footing

Zapata de columna:
Spot footing

Zapatilla eléctrica:
Power strip

Zona de no fumar:
No smoking area

CHAPTER 22/CAPITULO 22
English-Spanish Phrases

ENGLISH-SPANISH PHRASES
(ALPHABETICAL LISTING)

Alternate ties on this mat.	Alternar las ataduras en ésta estera.
Are you hungry?	¿Tiene hambre?
Are you hurt?	¿Está herido?
Are you thirsty?	¿Tiene sed?
Authorized personnel only.	Personal autorizado solamente.
Be careful.	Cuidar. Ten cuidado.
Be sure to call before you dig.	Asegurese de llamar antes de que usted cave.
Bring your safety glasses, hard hats and tools.	Traer sus gafas de seguridad, cascos, y herramientas.
Call for help.	Llama por ayuda.
Can someone interpret?	¿Puede alguien interpretar?
Can you drive a car?	¿Sabe conducir?
Can you rig and fly this joist?	¿Puede usted emparejar y volar esta vigueta?
Can you sign this ticket?	¿Puede usted firmar éste recibo?
Can you work tomorrow?	¿Puede trabajar mañana?

ENGLISH-SPANISH PHRASES
(ALPHABETICAL LISTING) *(cont.)*

Change the water in your bucket often.	Cambia el agua de tu cubeta a menudo.
Change your radio battery.	Carge su pila de radio.
Check the ground pin of the electrical plug.	Revisa la cuchilla a tierra de clavija eléctrica.
Check the plans.	Revisa los planos.
Check your contract.	Revisa tu contrato.
Clean and oil the forms.	Limpia y aceita las formas.
Clean this.	Limpie ésto.
Come with me.	Venga conmigo.
Do you have a driver's license?	¿Tiene licencia de conducir?
Do you have your own tools?	¿Tiene sus propias herramientas de mano?
Do you know crane hand signals?	¿Sabe usted señales manuales de grua?
Do you speak English?	¿Hablas Inglés?
Do you speak Spanish?	¿Hablas Español?
Do you use alcohol?	¿Usa usted alcohol?
Do you use drugs?	¿Usa usted drogas o narcótico?
Do you use medicine?	¿Usa usted alguna medicina?
Don't go in that area, it's secured for asbestos removal.	No entre en ésa area, ésta asegurada por el retiro de asbestos.

Don't look at the arc.	No mire al arco.
Don't move.	No se mueva.
Don't put too much in your wheelbarrow.	No ponga demasiado en su carretilla.
Double check.	Recomprobar. Revise nuevamente.
Drill a hole in the form.	Taladre un agujero en las formas.
Everything is ready.	Todo está listo.
Excuse me. (apology)	Disculpe.
Excuse me. (permission)	Con permiso.
Feet and inches	Pies y pulgadas
Figure it out.	Deduzcalo.
Finish up.	Terminar.
Get it done quickly and safely.	Hazlo rapidamente y con seguridad.
Get it done.	Cumplelo.
Get me the ___.	Traigame el (la) ___.
Get the first aid kit.	Traigame el juego de primeros auxilios.
Give me	Da me
Go for help.	Vaya por ayuda.
Good work.	Buen trabajo.
Hard hats and safety glass are required at all times.	El casco y las gafas de seguridad se requieren siempre.

ENGLISH-SPANISH PHRASES
(ALPHABETICAL LISTING) (cont.)

Hold the grade stick plumb.	Detenga el palo de grado plomado.
How do you say___?	¿Cómo se dice ___?
How many yards did you order?	¿Cuantas yardas oredeno usted?
How many?	¿Cuantos?
How much?	¿Cuanto?
Hurry up.	Apurarse.
I don't understand.	No comprendo.
I need	Necesito, necesita
I need more three-inch bolts.	Necesito mas tornillos de tres pulgadas de largo.
I need to see the identification you listed on the form.	Necesito ver la identificación que indico en el formulario.
I need you here early to open the gate.	Te necesito aqui temprano para que abras la puerta.
I need you to fill out this form.	Necesito que llenes éste formulario.
I'm sorry.	Lo siento.
Is the ladder secured at the top?	¿Está la escalera fija arriba?
Is the preparation all done?	¿Está la preparación toda hecha?
Is the welding lead connected well?	¿Está el cable conductor de soldadura conectado seguramente?

ENGLISH-SPANISH PHRASES
(ALPHABETICAL LISTING) *(cont.)*

Keep off the roof unless you are tied off.	No se suba al techo al menos que este atado.
Keep out.	Mantengase fuera.
Keep the job site clean.	Mantenga el trabajo limpio.
Keep your bead even.	Mantenga su cordon parejo.
Keep your trowel flatter.	Mantenga su llana mas aplanada.
Make sure the impalement protection is in place.	Asegurese que las protección del impalamiento este en su lugar.
Measure that stud again.	Mide ése montante otra vez.
Move this.	Mueva esto.
My name is ___.	Mi nombre es ___.
Never tamper with the guard on a Skill saw.	Nunca juegues con el protector del serrucho eléctrico.
No smoking.	No fumar.
Not ready yet.	No ésta listo todavia.
Pay attention to all warning signs.	Presta atención a todas las señales de peligro.
Pleased to meet you.	Mucho gusto.
Precisely (time)	En punto (tiempo)

ENGLISH-SPANISH PHRASES
(ALPHABETICAL LISTING) *(cont.)*

Put up the "Slippery When Wet" sign when you mop.	Ponga la señal "Resbaloso Cuando Mojado" cuando usted trapee.
Report all accidents to your supervisor immediately.	Reportar todos los accidentes a su supervisor inmediamente.
Salvage any metal that you find.	Salve caulquier metal que encuentre.
Snap a line.	Echar una raya línea.
Stand back while I break this glass.	Hagase atras mientras que rompa este cristal.
Stay clear.	Quedarse alejado.
Strike an arc.	Prender un arco.
Strip the forms.	Pelar las formas.
That is a damaged electrical cord.	Este es un cable eléctrico dañado.
That joint is still hot.	Este empalme esta todavia caliente.
That section is cut to grade.	Esta sección esta cortada a nivel.
The column has a loose connection.	La columna tiene la conexión floja.
The fire extinguisher should be kept with the cutting torch.	El extintor se debe guardar con la antorcha de corte.
The first aid kit is in the trailer.	El juego de primeros auxilios esta situado en el trailer.

ENGLISH-SPANISH PHRASES
(ALPHABETICAL LISTING) *(cont.)*

The floor has been waxed, you'll have to wait until it is dry to walk on it.	Se ha encerado el piso, tendra usted que esperar hasta que se seque antes de caminar en él.
The plans and specs don't agree.	Los planos y las especificaciónes no concuerdan.
The short blue flame is the hottest.	La llama azul corta es la mas caliente.
These one by eights are all warped.	Todos estos uno por ochos estan torcidos.
This cord doesn't have a ground prong.	Este cable eléctrico no tiene la cuchilla a tierra.
This ground is hard and solid.	Esta tierra es dura y solida.
This ground is muddy.	Esta tierra esta lodosa.
This is your responsibility.	Esta es su responsabilidad.
This wall isn't square.	Esta pared no esta cuadrada.
Use bracing if there could be a cave-in.	Utilice riostras si hay cualquier posibilidad de colapso.
Use screws instead of nails.	Utiliza tornillos en vez de clavos.
Use sweeping compound when you sweep the floor.	Utilice el compuesto de barrida cuando barra el piso.
Vibrate over here.	Vibre aquí.

Watch out.	Poner atáncion.
Watch out for protruding nails.	Fijese por clavos salientes.
Watch the chute.	Este atento con él canal inclinado.
Watch your fingers.	Cuidado con sus dedos.
We need another egress ladder.	Necesitamos otra escalera de salida.
Wear these dust masks.	Usen éstas mascarillas de polvo.
Wear this for your protection.	Use esto para su protección.
What does____ mean?	¿Que significa ___?
What is your name?	Como se llama?
What is your phone number?	¿Cual es tu número de teléfono?
What is?	¿Qué es?
What time?	¿Que hora es?
What will the width of this channel be?	¿Cual va a ser la anchura de este canal?
Where does it hurt?	¿Donde te duele?
Where is the water jug?	Donde esta la jarra de agua potable?
Where is your ___?	¿Donde esta su ___?
Will you deliver and lay the sod?	Puede entregar y poner el césped?

ENGLISH-SPANISH PHRASES
(ALPHABETICAL LISTING) *(cont.)*

Without ID, I can't hire you.

Sin la identificación adecuada, no puedo emplearlo.

Work safely.

Trabaje con cuidado.

You missed a spot.

Falto un lugar. Le faltoun lugar.

You're welcome.

Gracias.

Your pay will be ___ per hour.

Se le va a pagar ___ por hora.

... less tax withholdings.

... menos descuentos por impuestos.

NOTES

CHAPTER 23/CAPITULO 23
Español-Inglés Frases

ESPAÑOL–INGLÉS FRASES
(LISTADO ALFABÉTICO)

Español	Inglés
Alternar las ataduras en esta estera.	Alternate ties on this mat.
Apurarse.	Hurry up.
Asegurese de llamar antes de que usted cave.	Be sure to call before you dig.
Asegurese que la proección del impalamiento este en su lugar.	Make sure the impalement protection is in place.
Buen trabajo.	Good work.
Cambia el agua de tu cubeta a menudo.	Change the water in your bucket often.
Carge su pila de radio.	Change your radio battery.
Cumplelo. Hazlo.	Get it done.
¿Cómo se dice ___?	How do you say___?
¿Cómo se llama?	What is your name?
Con permiso.	Excuse me. (permission)
¿Cual es tu número de teléfono?	What is your phone number?
¿Cual va a ser la anchura de éste canal?	What will the width of this channel be?

Español	Inglés
¿Cuantas yardas oredeno usted?	How many yards did you order?
¿Cuanto?	How much?
¿Cuantos?	How many?
Cuidado con sus dedos.	Watch your fingers.
Da me.	Give me.
Deduzcalo.	Figure it out.
Detenga el palo de grado plomado.	Hold the grade stick plumb.
Disculpe.	Excuse me. (apology)
¿Donde esta la jarra de agua potable?	Where is the water jug?
¿Donde esta su ___?	Where is your ___?
¿Donde te duele?	Where does it hurt?
Echar una raya.	Snap a line.
El casco y las gafas de seguridad se requieren siempre.	Hard hats and safety glass are required at all times.
El extinor se debe guardar con la antorcha de corte.	The fire extinguisher should be kept with the cutting torch.
El juego de primeros auxilios estan situados en el trailer.	The first aid kit is in the trailer.
En punto (tiempo)	Precisely (time)

ESPAÑOL–INGLÉS FRASES
(LISTADO ALFABÉTICO) *(cont.)*

Ese empalme esta caliente todavia.	That joint is still hot.
¿Esta el cable conductor de soldadura conectado seguramente?	Is the welding lead connected well?
Esta es su responsabilidad.	This is your responsibility.
¿Esta herido?	Are you hurt?
¿Esta la escalera fjia arriba?	Is the ladder secured at the top?
¿Esta la preperación toda hecha?	Is the preparation all done?
Esta pared no ésta cuadrada.	This wall isn't square.
Esta tierra es dura y solida.	This ground is hard and solid.
Este atento con el canal inclinado.	Watch the chute.
Este cable eléctrico no tiene la cuchilla a tierra.	This cord doesn't have a ground prong.
Este es un cable eléctrico dañado.	That is a damaged electrical cord.
Esta sección esta cortado a nivel.	That section is cut to grade.
Esta tierra esta lodosa.	This ground is muddy.
Fijese por clavos salientes.	Watch out for protruding nails.

Gracias.	You're welcome.
¿Hablas Español?	Do you speak Spanish?
¿Hablas Inglés?	Do you speak English?
Hagase atras mientras que rompo éste cristal.	Stand back while I break this glass.
Hazlo rapidamente y con seguridad.	Get it done quickly and safely.
La columna tiene la conexión floja.	The column has a loose connection.
La llama azul corta es la mas caliente.	The short blue flame is the hottest.
Le falto un lagar.	You missed a spot.
Limpia y aceita las formas.	Clean and oil the forms.
Limpie ésto.	Clean this.
Llama ayuda.	Call for help.
Lo siento.	I'm sorry.
Los planos y las especificaciónes no concuerdan.	The plans and specs don't agree.
Mantenga el trabajo limpio.	Keep the job site clean.
Mantenga su cordon parejo.	Keep your bead even.
Mantenga su llana mas aplanada.	Keep your trowel flatter.

ESPAÑOL–INGLÉS FRASES
(LISTADO ALFABÉTICO) *(cont.)*

Mantengase fuera.	Keep out
Mi nombre es ___.	My name is ___.
Mide ése montante otra vez.	Measure that stud again.
Mucho gusto.	Pleased to meet you.
Mueva esto.	Move this.
Necesitamos otra escalera de salida.	We need another egress ladder.
Necesito mas tornillos de tres pulgadas de largo.	I need more three-inch bolts.
Necesito que llenes éste formulario.	I need you to fill out this form.
Necesito ver la identificación que indico en el formulario.	I need to see the identification you listed on the form.
Necesito, necesita.	I need.
No comprendo.	I don't understand.
No entre en esa área, esta asegurada por el retiro de asbestos.	Don't go in that area, it's secured for asbestos removal.
No esta listo todavia.	Not ready yet.
No fumar.	No smoking
No mire al arco.	Don't look at the arc.

No ponga demasiado en su carretilla.

Don't put too much in your wheelbarrow.

No se mueva.

Don't move.

No se suba al techo al menos que este atado.

Keep off the roof unless you are tied off.

Nunca juegues con el protector del serrucho eléctrico.

Never tamper with the guard on a Skill saw.

Pelar las formas

Strip the forms

Personal autorizado solamente.

Authorized personnel only

Pies y pulgadas.

Feet and inches.

Poner atánción.

Watch out

Ponga la señal "Resbaloso Cuando Mojado" cuando usted trapee.

Put up the "Slippery When Wet" sign when you mop.

Prender un arco.

Strike an arc.

Presta atención a todas las señales de peligro.

Pay attention to all warning signs.

¿Puede alguien interpretar?

Can someone interpret?

¿Puede entregar y poner el césped?

Will you deliver and lay the sod?

¿Puede trabajar mañana?

Can you work tomorrow?

¿Puede usted emparejar y volar este vigueta?	Can you rig and fly this joist?
¿Puede usted firmar este recibo?	Can you sign this ticket?
¿Qué es?	What is?
¿Que hora es?	What time?
¿Qué significa ___?	What does ___ mean?
Quedarse alejado.	Stay clear.
Recomprobar.	Double check.
Revisa la cuchilla a tierra de clavija eléctrica.	Check the ground pin of the electrical plug.
Revise los planos.	Check the plans.
Revise su contrato.	Check your contract.
Reporter todos los acidentos a su supervisor inmediamente.	Report all accidents to your supervisor immediately.
¿Sabe conducir?	Can you drive a car?
¿Sabe usted señales manuals de grua?	Do you know crane hand signals?
Salve caulquier metal que encuentre.	Salvage any metal that you find.
Se ha encerado el piso, tendra usted que esperar hasta que se seque antes de caminar en él.	The floor has been waxed, you'll have to wait until it is dry to walk on it.

ESPAÑOL–INGLÉS FRASES
(LISTADO ALFABÉTICO) *(cont.)*

Se le va a pagar ___ por
hora.... menos descuentos
por impuestos.

Your pay will be
___ per hour.... less
tax withholdings.

Sin la identificación adecuada,
no puedo emplearlo.

Without ID, I can't
hire you.

Taladre un agujero
en las formas.

Drill a hole in the form.

Te necesito aquí temprano
para que abras la puerta.

I need you here early
to open the gate.

Ten cuidado.

Be careful.

Terminar.

Finish up.

¿Tiene hambre?

Are you hungry?

¿Tiene licencia de conducir?

Do you have a
driver's license?

¿Tiene sed?

Are you thirsty?

¿Tiene sus propias
herramientas de mano?

Do you have your
own tools?

Todo ésta listo.

Everything is ready.

Todos éstos uno for
ochos estan torcidos.

These one by eights
are all warped.

Trabaje con cuidado.

Work safely.

Traer sus gafas de seguridad,
cascos, y herramientas.

Bring your safety glasses,
hard hats and tools.

Traigame el juego de primeros auxilios.	Get the first aid kit.
Triagame el (la) ___.	Get me the ___.
¿Usa usted alcohol?	Do you use alcohol?
¿Usa usted alguna medicina?	Do you use medicine?
¿Usa usted droga narcótico?	Do you use drugs?
Use esto para su protección.	Wear this for your protection.
Usen éstas mascarillas de polvo.	Wear these dust masks.
Utilice el compuesto de barrida cuando barra el piso.	Use sweeping compound when you sweep the floor.
Utilice riostras si hay cualquier posibilidad de colapso.	Use bracing if there could be a cave-in.
Utiliza tornillos en vez de clavos.	Use screws instead of nails.
Vaya por ayuda.	Go for help.
Venga conmigo.	Come with me.
Vibre aquí.	Vibrate over here.

About The Author

Paul Rosenberg has an extensive background in the construction, data, electrical, HVAC and plumbing trades. He is a leading voice in the electrical industry with years of experience from an apprentice to a project manager. Paul has written for all of the leading electrical and low voltage industry magazines and has authored more than 30 books.

In addition, he wrote the first standard for the installation of optical cables (ANSI-NEIS-301) and was awarded a patent for a power transmission module. Paul currently serves as contributing editor for *Power Outlet Magazine*, teaches for Iowa State University and works as a consultant and expert witness in legal cases. He speaks occasionally at industry events.

Acerca Del Autor

Paul Rosenberg posee un vasto conocimiento en las áreas de la construcción, la información, los sistemas eléctricos y de HVAC (sistemas de calefacción, ventilación y de aire acondicionado) y las instalaciones de cañerías. Es una voz destacada en la industria eléctrica con años de experiencia, que incluyen desde el desempeño como aprendiz hasta director de proyecto. Paul ha escrito para todas las revistas destacadas de la industria eléctrica y de baja tensión, y es el autor de más de 30 libros.

Además, escribió la primera norma para la instalación de cables ópticos (ANSI-NEIS-301) y recibió una patente por un módulo de transmisión de energía. Actualmente, se desempeña como redactor colaborador para la revista Power Outlet, enseña en la Universidad del estado de Iowa y trabaja como consultor y testigo experto en casos legales. Ocasionalmente, brinda disertaciones en eventos industriales.